T0178914

CLIMATE CHAOS AND ITS ORIGINS IN SLAVERY AND CAPITALISM

Climate Chaos and Its Origins in Slavery and Capitalism

REVA BLAU
AND JUDITH BLAU

ANTHEM PRESS

Anthem Press
An imprint of Wimbledon Publishing Company
www.anthempress.com

This edition first published in UK and USA 2020
by ANTHEM PRESS
75–76 Blackfriars Road, London SE1 8HA, UK
or PO Box 9779, London SW19 7ZG, UK
and
244 Madison Ave #116, New York, NY 10016, USA

British Library Cataloguing-in-Publication Data
A catalogue record for this book is available from the British Library.

Library of Congress Control Number: 2020941009

ISBN-13: 978-1-78527-527-2 (Hbk)
ISBN-10: 1-78527-527-5 (Hbk)

This title is also available as an e-book.

To Greta Thunberg and the Juliana Plaintiffs, and to Dashiell and Siena, the young people we love the most.

CONTENTS

PREFACE

Judith is the mother of Reva who is the mother to Dashiell age 15 and Siena age 8. Together, we wonder if it is even ethical for future generations to reproduce when children today will undoubtedly face significant challenges wrought by the climate crisis. The COVID-19 crisis underlines this existential dilemma even more starkly.

Like climate change, the pandemic requires the kind of global cooperation and solidarity that matches the speed of global, late-stage capitalism. Instead, countries were left on their own to respond to the crisis weeks after it had already taken hold in some of the most advanced societies of the world—China, Italy, Spain, and the United States. Within the United States, the far-right policies of Donald Trump prevented any type of preparation. In fact, while members of the Intelligence Community warned last year of a pandemic on the scale we are seeing today, the administration systematically cut off the epidemiology, which forewarned the crisis.

Notably, at the dawn of the crisis, Donald Trump systematically contradicted Dr. Anthony Fauci, the well-respected physician and immunologist, who emerged as the nation's expert on the control of infectious diseases.

The White House continues to silence and contradict epidemiologists, allowing states to chart their own course with social distancing protocols. In a bizarre Faustian bargain, states that were slow to test, slow to quarantine, slow to discourage travel and shopping are now being awarded with more medical equipment to respond to the spikes they are seeing in cases, hospitalization, and patients needing acute care. These are the same red states who voted for Trump creating a bizarre Faustian bargain with the president of the United States.

The Trump administration to this day has also refused to federalize the response to the crisis, even while, at the time of writing, 135,000 Americans have died, and well over three million people are infected. States and localities are left on their own to respond to the crisis by bidding against each other for ventilators and PPEs and begging for volunteers with nursing experience, either in school or retired, to join the ranks of healthcare professionals. The response has been so fragmented that even trace testing and reporting across the United States are titanic tasks.

Clearly and tragically, it is the death count that lights up a litmus test in orange neon the brutality of capitalism as a system for organizing society when faced with a pandemic. Countries with early, federalized coordinated responses have fared well. Those without such responses, largely due to the fears of the effects of social distancing on the marketplace, have not. As the *Washington Post* reported it, "as of April 8, South Korea had suffered 200 deaths due to the virus (4 per 1 million of population) and the number of new cases has slowed, while the United States had suffered 13,000 deaths (39 per 1 million population) with new cases continuing to grow quickly" (*Washington Post*, April 10, 2020).

Climate change presents the same challenge logistically and morally to a system created almost solely for profit. We have yet to arrive at the apex of the COVID-19 crisis. One can only hope that in the wake of this tragedy we will learn that we must work together to save lives.

Dash was born a year before Al Gore published his 2006 seminal book, *An Inconvenient Truth: The Planetary Emergency of Global Warming and What We Can Do about It*. Having taken a freshman class on Earth Science with Wallace Broecker at Columbia, I (Reva) knew of human-caused climate change in the late 80s. I remember distinctly the graphs he had created showing the climate fluctuations from the Pleistocene era. I also remember distinctly the more recent graphs and the point at which (around 1980) CO_2 emissions and average global temperatures began their steep, unfathomable ascent in red pen, creating an upward arc that seemed, even then, to risk ascending ever upward toward a vertical future off the graph.

Yet, I still viewed climate change, and even its catastrophic effects, as unfolding somewhere out there beyond the confines of my existence in Paris, New York, and Massachusetts. It was the stuff of science and it occurred in nature, which I saw as unfolding well outside my life on the Upper West side and in various other places. I felt greater fears, I must confess, for the flora and fauna, which I presumed would suffer the most.

Reading Gore's book when Dash was just a year old awakened in me the sense that climate change could be a crisis. I probably would have been keener in producing another child if it had not been for Gore's book and movie. Instead, Dash's dad, Joe, and I decided to adopt a child. Some people thought that the concept of "recycling" was a strange way of coming into parenting; but I thought it made all the sense in the world.

I also am a middle school teacher. The Monday after the Climate Strike, a sixth-grade student raised her hand and, echoing Greta Thunberg's speech at the UN, said, "We have eight and a half years to halt climate change and save ourselves. Do you think we will?" She was referring to the amount of emissions that could still be emitted without passing the 1.5 degree limit. At the current rate of emissions, the world is predicted to reach it in eight-and-a-half years. I had to be honest with her and I said, "I don't know."

But I do hope that this book, if it does anything, contributes to the enormous push—primarily from people between her age and the age I was when I first saw Wallace Broecker's graphs on the projector, to demand from governments the type of changes that could, indeed, halt climate change and avoid the most disastrous effects from wreaking suffering, or even death, of the billions of people who call Earth their home.

Global warming is happening faster and with more devastating consequences than predicted even a year ago. We are entering a true carbon-fueled crisis, one that will have devastating effects on all future generations and could well make the Earth permanently uninhabitable by the century's end. By the time today's children are adults, it is almost a given that the world will look completely different than today, and it will be reshaped by climate change. By the time of our grandchildren's adulthood, the Earth very well might be unlivable.

In December 2015, the Paris Agreement, named after the city in which it was adopted, brought 186 countries together to strengthen the global response to climate change. More ambitious than its predecessors, the Paris Agreement's goal was to keep warming to well below two degrees Celsius, which would halve the trajectory of warming that scientists predict through the projection models using the rate of emissions today. Initially the world's biggest emitters, China and the United States, who had not participated in the Kyoto Protocol, joined the Agreement to significantly reduce emissions, and help developing countries reduce theirs. However, when Trump took office, he announced that the United States would withdraw. That decision had—and continues to have—cruel and global significance since the United States has such high carbon dioxide emissions that it harms the world's peoples. Judith had written a hopeful book—*The Paris Agreement: Climate Change, Solidarity, and Human*—which was published in 2017. Yet one of the first things Trump did when becoming president of the United States was to withdraw from the Paris Agreement. The world's scientists were horrified, knowing the consequences for the planet and the people who inhabit it. In response, Judith wrote and published, *Crimes against Humanity: Climate Change and Trump's Legacy of Planetary Destruction* (2019).

Today, scientists are concerned that the Paris Agreement did not go far enough or that countries are not meeting their targets. Even by limiting warming to one-and-a-half degrees from pre-industrial levels, the goals of the Paris Agreement, human civilization will be gravely threatened by extreme heat waves and drought, pestilence, fires, and species extinction. A band around the Earth closest to the equator will become unlivable—sending millions of refugees to resettle. The Arctic, Greenland, and Antarctica's ice sheets might still reach some key tipping points, with melting ice sending methane into the air, causing sea-level rise of several feet, exposing 30 to 80 million people to coastal flooding and disrupting ocean and atmospheric circulation.

Warming has occurred since 1975, at a rate of roughly 0.15–0.20°C per decade, and people contend with storms, extreme heat, and extreme weather events, along with massive immigration and civil wars, including genocide, that the heat and its concomitant food and water insecurity extreme heat catalyzes. After the Paris Agreement, global carbon dioxide emissions skyrocketed, leading some scientists to predict that we are not on the path for four degrees of warming; in fact, we are on the path for six degrees. While we tend to believe that the experts are not doing enough to help solve the problem directly, in fact, they are engaged in collaboration, discovery, education, and coordination while working with governments, NGOs, corporations, and communities. All this involves building trust, collaboration, and information exchange.

The deadline is 2025 for the end of support for fossil fuels and the deadline for phasing in of 100 percent renewable energy is 2050. This process and the deadlines are critical. What we do today, or indeed in the next five or ten years, will enormously affect the future to mitigate the disastrous consequences of our prior myopia. The structural change is enormous, but it is one that, as we reveal, has been at the heart of the class and colonial struggle since the very beginnings of the modern era. The financial system of capitalism which, like a god run amuck, has wreaked havoc on that Goldilocks levels of oxygen, water, and temperance that make the planet safe for humans also is failing in a myriad of other ways, including creating apartheid levels of wealth inequality, government corruption, and rigged elections. No longer seen as an unfortunate byproduct of free markets, wealth inequality has become a veritable caste system whereby the 1 percent extracts so systematically from the rest of us that even the ideas of "free" and "market" are increasingly contradictory. The free market, once seen as salutary, has become exposed as the engine through which humans spew the very poisonous gases into the air that will choke us and cause our extinction.

Millions of activists around the world have called for an immediate end to burning fossil fuels. One of the most successful movements—Extinction Rebellion, which started in London, and is gaining momentum worldwide, suggests the five ways that have shifted since environmental movements of the previous generations. First, it is comprised of young people outside of any formal organization. Second, it plainly demands for both scientists and policy makers, including national governments, to "tell the truth." As they outline it, the truth is that the climate crisis spells catastrophe in the next decade and could result, if not, in the extinction of the human species by the century's end, a crossing of a threshold of tipping points whose interactions could greatly accelerate the end of civilization.

Third, heeding the work of climate change scientists, the Extinction Rebellion as well as the Fridays for Future Movement led by Greta Thunberg, and *Juliana et al. vs. United States of America* are climate justice actions that seek to move both scientific and policy discourse around climate change. They point out that science has been reticent to push policy on climate, since it has generally embraced what is called the path of "least drama." Even the hundreds of reports synthesized by the Intergovernmental Panel on Climate Change (IPCC) understate the severity, urgency, and interrelatedness of anthropogenic changes due to reticence in both science and also in policy recommendations. Fourth, it seeks to restore the globe to both environmental and economic justice simultaneously by joining the rights of migrants, indigenous peoples, and the disenfranchised to the fate of human society. Fifth, Extinction Rebellion encourages people to take radical, nonviolent action to disrupt the capitalist machine. Finally, it demands that Citizens' Assembly on climate and ecological justice lead policy makers, who they see as beholden to the fossil fuel industry, to ending fossil fuel extraction.

There is no doubt that the climate emergency that scientists and, even some leaders are beginning to acknowledge. The global pandemic highlights the consequences of ignoring the cascading and multi-interactional effects of further destroying ecosystems and habitats of key species that keep the Earth in balance. Yet the climate crisis mandates us to remake even a physically diminished world in the image of indigenous societies, centering the natural world alongside human rights, and truly valuing sustainability and equity, in accordance with abolitionist and decolonization movements. Unlike previous models, which have seen developing countries as a problem, the new one sees the less industrialized societies as providing us with the blueprint for sustainability. We view this re-centering of Mother Earth and the dismantling of the capitalist system as no longer a utopian vision of the colonial struggle, but a prerequisite for human survival.

A strength of this book is that it clarifies that capitalism is the culprit for global warming. We also have a chapter on what the UN is considering as an alternative to capitalism, namely, the circular economy. We provide a clear explanation as to how slavery primogenitured American's extreme and brutal version of capitalism and set in motion the machinery of climate crisis, which amplifies the human tragedy with which capitalism began. We have a chance at creating a better world based on systems that put humans as ecological and ethical animals with advanced systems of self-knowledge in the center of our decisions. We must seize it before it too late.

CHAPTER 1

BACKGROUND: EARLY SIGNS OF WARMING INTO THE PRESENT

It is particularly poignant that climate science owes its genesis to two unusually cold summers in the Alps two centuries ago. After the "year without a summer" that stretched into two frigid years in the Alps, the Swiss Natural Science Foundation formed a competition to address the aberrant climate event. Ignaz Venetz, an engineer from the Valais, won the competition with his research and observational paper "Mémoire sur les Variations de la température dans les Alpes de la Suisse." In it, Venetz used painstaking observational evidence and oral history of local villagers to come up with the theory of glacial advance and retreat, breaking with the long-standing scientific consensus that climate was unvarying.

A trekker in the majestic alpine terrain and a native of the Valais, the highest canton of Switzerland, Venetz described the moraines in which he saw evidence of lost glacials in great, sometimes even sumptuous, detail. Speaking of le lac de Champée surrounded by the noble glaciers, he writes:

Il est impossible de résister à cette magie du sentiment qu'inspire la vue d'un spectacle si extraordinaire. En vain tenteroit-on de peindre ce que l'on éprouve sur une scène si pittoresque et majestueuse où se présentent un grand nombre de cimes aériennes groupées autour de ces géans des Alpes, qui tantôt portent leurs fronts audacieux jusques dans les sombres nuées, tantôt découvrent leur tête couronnée de mille rayons, dont l'éclat, rehaussé par le reflet de glace, transporte l'âme en la remplissant des charmes les plus doux. Si l'effet de ce coup-d'oeil est si prodigieux même sur l'habitant des alpes, accoutumé à voir la nature dans toute sa majesté, quel ne doit pas être ravissement du citadin ou de celui qui, élevé loin des montagnes, n'a jamais rien contemplé de semblable?[1]

---

1 Ignaz Venetz, Mémoire sur les variations de la température dans les Alpes de la Suisse, rerodoc. Digital library, 1621: http://doc.rero.ch/record/19750 p. 3.

1

[It is impossible to resist the magic evoked by the view of such an extraordinary spectacle. In vain, would one paint what one experiences from a scene so picturesque and majestical which presents a large number of aerial peaks grouped around the giants of the Alps, which sometimes wear their audacious faces into the dark clouds, sometimes unveiling their head crowned by a thousand rays of lights, whose bursts, enhanced by the mirror reflection, transports the soul while filing it with the sweetest gifts. If the effect of this glance is so rich even to an inhabitant of the Alps, accustomed to seeing Nature in all its majesty, how ravishing it must be for town dweller or one who, raised far from the mountains, who has never contemplated something similar? (Translation by author)]

In this case, in a bitter irony, the science itself shows also the human and aesthetic sense of what will be lost with climate change. Tellingly, in Switzerland this year locals marked the loss of the Pizol glacier in the Glarus Alps, eastern Switzerland, which has in the last 10 years lost 90 percent of its mass, with a funeral. Hundreds of people dressed in funeral black performed a ceremony marking the death.

The greenhouse effect was outlined a few years later, when French physicist Joseph Fourier discovered that the composition of the atmosphere contributes to the warming of the planet.[2] In 1861, Irish physicist John Tyndall identified carbon dioxide as a gas that would contribute to the greenhouse effect and went all over London and abroad to publicize his findings.[3]

It was not until the end of the nineteenth century that scientists in Sweden, the world's first physical chemists, would mathematically demonstrate that human behavior, namely the burning of coal and fossil fuels to drive the development of advanced countries, contributed to the warming of the Earth. Svante Arrhenius won the Nobel Prize for Chemistry in 1903, for his work in the 1890s with colleague Arvid Hogbom in calculating the causal relationship between CO_2, including human-caused CO_2, and increases in the Earth's surface temperature.[4]

---

2 Danny Lewis, "SmartNews: Scientists Have Been Talking about Greenhouse Gases 191 Years." *Smithsonian Magazine*: https://www.smithsonianmag.com/smart-news/scientists-talking-about-greenhouse-gases-191-years-180956146/.

3 "The Discovery of Global Warming: The Carbon Dioxide Greenhouse Effect," January 2020: https://history.aip.org/climate/co2.htm.

4 Svante August Arrhenius, "The Nobel Prize in Chemistry, 1903": https://www.nobelprize.org/prizes/chemistry/1903/summary/; "Climate Change Science: A Very Short History," Nottingham Science Blog: https://nottinghamscience.blogspot.com/2016/01/climate-change-science-very-short.html.

This advance formed the first paradigm shift in humans' role in nature as it existed in Western countries' philosophies. For the first time, nature was not in binary relationship with Earth—either a driving force to be conquered or transcended, or at the least, categorically separate from civilization; but that nature and humans were part of the same system. Like all scientific advances of the nineteenth century, however, these turn-of-the-century scientists still formed a part of the enlightenment mindset that believed, with teleological certainty, human history is bent toward greater and greater reason and advancement. The scientists who theorized climate warming, and the role of human behavior in it, for example, had faith that the oceans would absorb this extra CO_2. It would be a long time before scientists turned their attention to studying ocean absorption.

This faith, as it turns out, would be the precursor of any number of ideas that range from faith in the planet's natural processes to fix climate change, faith in technology to fix climate change, to outright climate denialism. This has only recently intensified in the United States, perhaps because of Trump's withdrawal from the Paris Agreement.

Even the oft-cited bias in science toward neutrality, objectivity, and against alarmism can be read now with nostalgia for a time when people, whether traditionally left or right, had faith in human history as progress. Note that this faith in the Earth, this idea that the Earth would simply go on as before, existed alongside theories, such as the various forms of manifest destiny and virulent systems of white supremacy, which legitimized wholesale the colonization of less powerful people by the more powerful.

From this perspective, the worst of the system we have inherited seems genocidal at its worst moments of human hubris and suicidal even when that very hubris contributed to all the wonder and joy we have created. It is hard to fathom that in the course of human history, the victors of world wars, and the countries that generated the most advanced universities, somehow missed the crucial and most simple of facts that the Earth's exalted Goldilock status—that combination of oxygen, water, and a temperate climate—was not immutable.

Looking back on the early modern era, we see that the underlying faith most people had toward the Earth and its flora and fauna was really the faith a spoiled child has toward a too-benevolent parent. This human blight by the way encompasses the world's worst genocidal dictators and also you and I, who still drive cars, take flights, shop at grocery stores, and use the internet.

The grandiose child that is featured annually at our family's Seder is a metaphor for human hubris, as it is been exercised since the age of colonialism and Industrial Revolution in Western societies. In the Haggadah the spoiled or grandiose child shows up as a cautionary tale for children reading the questions.

He is described as an "outcast" or as "lonely," which seems apt given that the very recent time frame in which humans had the time to act to save the planet's weather systems also accompanies a narcissism so exaggerated that it has become a caricature. This narcissism appears in the form of corporate billionaires continuing to burn fossil fuels at a rate thousands of times faster than the average person. In the Haggadah sometimes the spoiled child attempts to conquer the parent, as colonizers conquered native lands, pillaging and murdering people along the way; sometimes the child attempts to transcend the parent, as the developers destroyed forests to pave over in asphalt; and often the child simply ignores the parent, as most of us city and suburban dwellers and even the weekend hikers do, having the faith that the parent will always be there steadfast throughout.

We will be reaping the consequences.

If we listen to scientists today, we are about to enter an era that will be wholly defined by the anthropogenic heating of the Earth. A full half of the carbon emissions that we have emitted from the burning of fossil fuels in all of time has been spewed into the atmosphere since Al Gore's book *The Inconvenient Truth* was published.[5] In fact, the majority of the burning has happened within three decades, and 85 percent since World War II, which means that the story of human self-destruction is the story of the baby-boomer generation of human beings, including the lifetime of one of the co-authors of this book. That this lifetime also saw the rise of increased productivity and prosperity for all kinds of people, including minorities, women, and gay Americans, is one of the great and tragic ironies of the story of climate change.

Warming is measured by degrees difference from the pre-industrial era. The most exhaustive and also conservative reports are compiled by the United Nations' Intergovernmental Panel on Climate Change (IPCC; that only presents research which has been proven beyond a doubt). The IPCC measures temperature increase from the era 1850–1900. In 2017, scientists measure a one degree increase from pre-industrial levels. The best measurements for temperature rise, given current emissions, come from the Climate Tracker a climate science consortium that compiles data on climate science and tracks

---

5 Michelle Nijhuis, "What's Missing from an 'Inconvenient Sequel': Al Gore's New Climate Change Documentary?" *The New Yorker,* July 29, 2017: https://www. newyorker.com/science/elements/whats-missing-from-an-inconvenient-sequel-al-gores-new-climate-change-documentary.

both the science and the policy aimed at reducing emissions. On the "business-as-usual" scenario the Climate Tracker reports:

> In the absence of policies, global warming is expected to reach 4.1°C–4.8°C above pre-industrial by the end of the century. The emissions that drive this warming are often called Baseline scenarios [. . .] and are taken from the IPCC AR5 Working Group III. Current policies presently in place around the world are projected to reduce baseline emissions and result in about 3.3°C 1 warming above pre-industrial levels. The unconditional pledges and targets that governments have made, including NDCs 2 as of December 2018, would limit warming to about 2.8°C 3 above pre-industrial levels, or in probabilistic terms, likely (66% or greater chance) limit warming below 3.0°C.[6]

Should we continue down the same path, the IPCC tells us that we would see more than four degrees of warming. The impacts are unimaginable: fires in the American West would increase by a factor of 15–20, Europe would be in semi-permanent drought, sizable regions of the planet will be unlivable by direct heat, half of the world's capitals would be inundated with water, crop yields of staples such as corn and legumes would be cut by half, and more than a billion people would be climate refugees.

The UN IPCC predicts that at two degrees warming, the ice sheets will begin their collapse and so the coral reefs could well die off entirely, setting off a cascade of effects outlined above. We have probably already emitted enough carbon emissions in the atmosphere for half of that warming; so even if we reverse course today, we have already prescribed a very different climate from that of the previous generation. This new era will be more than a paradigm shift: it will flip 180 degrees the notion that history of human civilization, while ascending and descending peaks and valleys along the way, is always over time advancing toward something better. Many climate scientists have come out to say that the IPCC's reports are far too conservative. To take one example, the IPCC used a climate model to predict loss of Arctic ice which, when compared with recent years' modeling, is far too optimistic. While it showed a decline of ice that sloped downward toward the year 2100, recent observations have shown

---

6 Climate Action Tracker. "Pledge Pathways":　　https://climateactiontracker.org/methodology/pledge-pathways/.

a precipitous decline in just the last summer. This red line was well beneath the zone of probability that the IPCC reports had modeled.

Already, we are in the midst of the sixth mass extinction, or the Holocene extinction. It is not surprising that in this new era of mass extinction (the first five happened before recorded history), scientists would begin to question its native hostility towards passion and activism and begin to talk about the ethics of not being alarmist enough.

Climate experts, such as Naomi Orestes, Sir Nicholas Stern, and James Hansen, all point to several reasons that science, and most particularly attempts such as by the IPCC, has failed to keep apace of the climate emergency. One is that probability projections break down in the face of the complexity of interactions between human society and the multiple variables involved in temperature change and its outcomes. Will humans reduce their emissions or increase them? The only time during the last two decades that emissions decreased was during the economic downturn following the financial crisis. Quite sobering, since the IPCC report, emissions have skyrocketed, confusing the relationship to intentionality around reducing emissions in the face of facts.

The second is that the feedback loops involved as the poles melt (and the ice-free summers grow longer and threaten to become ice-free year round) are multiple, interactive, and unpredictable. Feedback loops create vicious circles that interact with each other and interact with humanity's attempt to survive them. For example, in simple terms, permafrost melt in the Arctic will trigger release of methane, a potent greenhouse gas, perhaps on the order of hundreds of gigatons, which both contributes to the trapping of heat in the atmosphere and, simultaneously, further reduces the albedo effect leading to even greater warming of the oceans, which then slows the currents that cool regions of the world.

The third, and most compelling, is that climate change collapses the difference between the cold objectivity demanded historically of scientists and the hot tempers required of political activists. The IPCC relied on reports from thousands of scientists each of whose work was read and scrutinized for three years by a vast network of scientists. Many scientists have now concluded that we simply do not have the luxury of that kind of time. If a study on the loss of Arctic ice, for example, would occur today, it would find that all the previous modeling was far too conservative. Yet, how useful would that be if by the time the new observations were released to the public, the loss of ice multiplies making these newer observations grossly outdated? Climate change is simply occurring at a faster rate than the standards that scientists have used can effectively capture.

Whereas science has depended on peer-review and the repetition of the experiment, it has never arrived at a model for human extinction. As the Extinction Rebellion points out, it is not the time to be analyzing how to avoid crashing into an iceberg, after the boat has crashed into the iceberg. By its very nature, science cannot account for disaster.[7]

It does not take a scientist to figure out what a mass extinction of animal species would mean for the animal kingdom's most advanced species. Ocean acidification caused by the absorption of carbon by the ocean has already led to an existential threat to the ocean life that is the basis of the food chain. Meanwhile, the destruction of the Earth's rain forest threatens the viability of land-based biodiversity, which compounds the disaster occurring in the oceans. According to a study published by the peer-reviewed journal, *Proceedings of the National Academy of Sciences*, the researchers Gerardo Ceballos, Paul R. Ehrlich, and Rodolfo Dirzo write, "Dwindling population sizes and range shrinkages amount to a massive anthropogenic erosion of biodiversity and the ecosystem services essential to civilization." By "civilization" the authors do mean human civilization, since such a massive loss of biodiversity, extending to animals not considered endangered, leads to their prediction that "humanity will eventually pay a very high price for the decimation of the only assemblage of life that we know in the universe."[8]

Just as we look toward past notions about human's relationship to nature as being as naive as Ansel Adams's pictures of the West, we can also look to counter-narratives to the Era of the Enlightenment and Conquest and the ancient or "primitive" cultural philosophies of the ancient world that valued Earth as a deity as being the objective narrative all along. In all forms of climate change, Noacine flooding, desertification, and the melting of the poles, one can see how much we have angered the gods. Whether they be in the form of the constitutions of countries like Bolivia which features both anti-colonialism and the inheritance of Mother Earth in its opening pages, hippies and hermits who decided in the 60s to go off grid, LCD-taking poets who renounced worldly possessions, or the spiritual traditions of any number of world cultures, all of them come into focus these days as being a heck of a lot wiser and more

---

7　Extinction Rebellion: https://rebellion.earth/the-truth/.

8　Gerardo Ceballos, Paul R. Ehrlich, and Rodolfo Dirzo, "Biological Annihilation via the Ongoing Sixth Mass Extinction Signaled by Vertebrate Population Losses and Declines." *Proceedings of the National Academy of Sciences* 114 (30). July 25, 2017: https://www.pnas.org/content/114/30/E6089.

pragmatic than the system driven by solipsistic heads of capital leading most of humanity on a suicide mission.

It would be naive to assume that all conquered people would have been better stewards of the planet but certainly conquered people have the humility that is necessary right now. There is plenty of evidence suggesting that the counter-Enlightenment philosophies that reinforce ideals of collaboration, solidarity, and shared opportunities and resources would have been better philosophies than the zero-sum game that climate change makes of resource competition in the form of advanced capitalism.

The geopolitical map is already being reshaped and not in the direction that the Paris Agreement would have favored. First, the global south, particularly in the Middle East and in parts of Africa, has seen an uptick in violence and full-scale civil war. These tides of violence rising as they do with extremist dictators thrive, not in darkness, but in extreme heat. While factors leading to war are complex, the rising heat and concomitant drought make food and water resources scarcer in regions, such as South Sudan and Syria, already prone to resource scarcity and resource conflicts. Ideological extremism flourishes when there is a lack of basic resources, easy access to weaponry, low education, and where, on top of those desperate state of affairs, it is too hot to grow arable crops. Add AK-47s to undereducated and desperate young men who can't feed their families; it is a cocktail for civil war and chaos. From the Syrian Civil War alone, there were 5.6 million Syrians who fled the country and another 6.2 million displaced. Half of the pre-war population has been uprooted and more than half required humanitarian assistance.

In reacting to the upsurge of migrants from the global south that roils with the perturbations both directly and indirectly by weather, many countries' governments have welcomed thousands of immigrants. But these efforts now have faced enormous backlash recently in the rise of nationalism and extremism across the West, both in European Union countries and America. In response to large-scale immigration, Europe, Great Britain, and the United States have been roiled by fascist or neo-fascist movements that threaten the very future of democracy. This surfacing of fascism threatens to eclipse the existential threat of climate change by keeping people in perpetual states of fear, division, and chaos. Corporate capitalism is becoming increasingly corrupt with corrupt influence over governments. An immediate example is traders using knowledge of Donald Trump's meetings with foreign leaders to make millions the day Trump tweets about a tariff. This blatant corruption makes it less possible for electorates to hold elected officials accountable to holding these white-collar criminals accountable.

The future remains to be written. This book lays out the possibility of how to respond to the climate crisis and it cascade of crisis, such as deadly global pandamics by suggesting the paradigm shift that must occur in considering the human role in nature. This is not a policy book, but a philosophical one. Without a modus operandi, the most privileged humans will continue behaving the way they always have, naively believing or worse, spreading the mythology of human progress against an eternal backdrop of nature when you need it, the abundance of clean water, oxygen, and food always there as a given. That clean water, oxygen, and food supply are no longer a given is creating such a myriad of problems, it's difficult to even fathom. It's a paradox for humans to think themselves out of the predicted consequences of the climate crisis in the future because without oxygen, water, and food, the ability to form sentences start to break down rather quickly.

We believe that climate change is the outcome of capitalism gone amuck but the end of capitalism, or its last gasp, does not come as a surprise to those of its history who were treated just as the Earth's other animals—chattel to be yoked, obstacles to clearing a land for settlement. To deal with climate change, we need first to recognize that capitalism was flawed from conception, and we need no further proof than the kidnapping, displacement, slavery, and genocide that accompanied the brutal and draconian settlement and reorganization of most of the Earth's populated surface during the sixteenth through twentieth centuries.

Attempts to geoengineer, like creating biodegradable plastic or putting the word *natural* on a label, try to make combating climate change palatable for the machinery of profit to churn on. Yet, there is no palatable way here and eventually, even the richest will feel the climate's wrath. Entire systems will need to transform fundamentally not only in the wealthiest countries, but those countries and tribal nations whose development and even well-being were mangled and thwarted for centuries. To confront climate change, we must humble ourselves by reflecting on capitalism's first historical sins borne of the centuries of empire formations—slavery and genocide. Second, we need to analyze the system whereby the next centuries' elites continued to extract infinite wealth and power by creating a caste society of consumers, poor whites, just slightly more advantaged than their black counterparts, in order to accumulate unparalleled wealth in the hands of a few. Today, we see the shamelessness with which these elites operate: by defunding education; interfering with elections; profiting off incarceration, disease, and addiction. To put it another way, we address climate change as a crisis embedded functionally and culturally with market systems of exchange. The reckoning of capitalism's horrific origins in slavery and its creation of an inhumane caste system reveal the contradictions of a political philosophy rooted in individualism, which nonetheless cares little for the human society, real human beings, and rather its own profitable ends by any means. Why would we trust this system with geoengineering?

Others have recognized that the impacts of climate change, heat, drought, famine, flooding, upending of militaries, millions of people displaced will be so massive that they will inevitably reshape the geopolitical world perhaps by the end of this century and certainly by the end of the next even if we stay within two degrees of warming. The possibilities of the political forms that will be built in the wake of its storms have been put forth in a book aptly called *Climate Leviathan*, whose authors Geoff Mann and Joel Wainwright sketch out four scenarios that play out along the lines of power and political ideas reformulated with the wholly new paradigm of a planet out of whack.[9]

In this book, we suggest a different model that gleams from liberal democracies: the lesson that diversity, democracy, and expertise make us all stronger. Yet it applies these lessons to a society stitched together with radically democratic local communities whose pride of place and sense of connection have these communities protect their environments much like a child protects its aging and infirm parent or a worshipper making tremendous sacrifices for the good of nature, as if she were a mother or a God. In this new political philosophy, humans are decentered and take their rightful place alongside the animal and plant kingdoms whose rights are considered inviolable. It will take much to re-balance the Earth, and it will take enormous ingenuity. But unless we reconceive the human will as that which strives toward good for all people beyond their function to create wealth, we won't be able to overcome our worst demons.

---

9 Wainwright, Joel, and Geoff Mann. *Climate Leviathan: a Political Theory of Our Planetary Future*. London: Verso, 2020.

CHAPTER 2

EXTRACTION: SLAVERY
AND CAPITALISM

Significant scholarship recently has focused on capitalism's origins in slavery
and white supremacy. One of the significant contributions to the literature came
in the form of an interactive project by the *New York Times*, "The 1619 Project,"
bringing the legacy of slavery in America into cars, homes, and workplaces
around the world. "In order to understand the brutality of American capitalism,
you have to start on the plantation," begins Episode 2 of the podcast. It is this
brutality enshrined in American history that paved the way for capitalism's
relentless brutalization of the environment.

From the beginning of the colonies, cotton became the driver of the new
economy and quickly became the commodity at the heart of the genesis,
the ethos, and the warped, predatory genius of capitalism. Cotton needed
enormous amounts of land, which was relatively plentiful in the New World.
What the production of cotton also needed was enormous amounts of labor to
pick and process the cotton. When we ask ourselves how humanity has ignored
the values of the Earth beyond its surplus value, we need to go back to slavery to
learn that this transactional and extractive ethos was born in the brutalization
of human beings for profit.

The invention of the cotton gin in 1794 by Eli Whitney caused the cotton
market to explode fueling the slave trade. "In 1790, we had just shy of about
700,000 enslaved workers on these shores. By 1850, that number is three million
enslaved workers in America, and cotton is driving most of that growth,"
writes Mathew Desmond.[1] To feed the hunger for profits, plantation owners
acquired more land or abandoned farms to find fertile soil as they bought

---

1 Mathew Desmond, "The Economy That Slavery Built," 1619 Project: https://www.
nytimes.com/2019/08/30/podcasts/1619-slavery-cotton-capitalism.html.

more slaves. They formed elaborate hierarchies underneath their ownership with an extensive network of kin and in-laws forming mid-level managers and overseers. If a slave resisted submission to brutalizing conditions, they were loaned out to poor whites who aspired to slavery's promises of riches. To this group, memorably, Ta-Nehisi Coates gives the namesake "Low Whites" in his novel, *The Water Dancers*. In history, "low whites," formed one bedrock of slavery, from the vigilante slave hunters who captured fugitives to the slave-less whites who exchanged their violent services as breaker of slaves for a year or more of labor on their land. Once the slave was broken, he would be returned to his owner. Low whites were the lynchpin onto which the hinges of slavery levered its great power. Since low whites enjoyed privileges and legal status as whites, they identified with elites, and were often willing to maintain the hierarchy of power created in slavery that functioned to put money and power into the hands of white elites, a stratum of money and power low whites would never themselves attain.

In Frederick Douglass's famous *Narrative of a Slave*, a student can count dozens of interlacing methods through which white supremacy enforced slavery. First and foremost, of course, was violence or threat of violence through inhuman punishment for any infraction, real or perceived. Whipping family members to punish an infraction or perceived disobedience was commonplace as was whipping to break the spirit. After his time with the Auld family in Baltimore, Douglass describes life as a field slave with his new master. Master Thomas tied a crippled woman to a tree and beat her over the course of a day while quoting the Scripture, "He that knoweth his master's will, and doeth it not, shall be beaten with many stripes." Since she had damaged hands, she was not a productive field hand. After beating her, he then sends her away to almost certain death. When Anthony decides that Douglass is not subservient enough, Anthony sends him to a tenant farmer notorious for breaking slaves in exchange for their labor. After being beaten almost to death, Douglass makes in vain a petition for protection leading him to standing up to Covey, the famous turning point in which Douglass becomes an abolitionist thinker.

Added to violence was ripping babies away from their mothers at birth, separating families at the auction block, withholding water and food, heat. Slave narratives record forced drunkenness at holiday time, which ensured that slaves felt hungover and ill during their rare days of furlough, longing to return to the normalcy of the field. The legal system in slavery which denied slaves the right to a trial, by definition, offered whites a blank check to inflict violence and immunity from any crime against blacks—free or enslaved. This immunity from responsibility is enshrined in America's earliest capitalist enterprise and continued through the crimes committed by gas companies who knowingly created a human disaster in the form of unlimited extraction of oil.

From the 1680s on, the legal system created categories for white men that allowed white men to enjoy the rights to property, including slaves, legal representation, and jury by peers. Meanwhile anti-miscegenation and property laws allowed for white men to marry white women while simultaneously accumulating property through the institutionalized rape of female slaves. Frederick Douglass writes,

> The whisper that my master was my father, may or may not be true; and, true or false, it is of but little consequence to my purpose whilst the fact remains, in all its glaring odiousness, that slaveholders have ordained, and by law, established, that the children of slave women shall in all cases follow the condition of their mothers; and this is done too obviously to administer to their own lusts, and make gratification of their wicked desires profitable, as well as pleasurable; for by this cunning arrangement, the slaveholder, in cases not a few, sustains to his slaves the double relation of master and father.[2]

While anti-miscegenation laws forbade mixed-race relations, white men were never held to these standards in a court of law, while they were unevenly and viciously upheld in the case of black men in relation with white women. (Such men would rarely survive and the women never allowed back into the community.) The property laws ensured that the progeny of slave owners and their slaves became property of the slave, incentivizing white men to rape black women.

Some of the most vicious apexes of slavery's history took place as slaves were sold when land was depleted by nutrient-ravaging cotton, just as today's most alarming environmental crimes against indigenous people are committed because of the declining fossil fuel resources that can be traditionally extracted. First-person narratives, such as Frederick Douglass's, reveal the expediency with which babies were separated from mothers to be sent to the elderly to be raised and then bought, sold, exchanged, or loaned depending on the financial needs of plantation owners linked through bonds of white supremacy.

When plantation owners expanded their farms, they took out mortgages to buy more land. Increasingly, after the Revolution, mortgages were not based on

---

2 Douglass, Frederick, and Angela Y. Davis. *Narrative of the Life of Frederick Douglass, an American Slave,* Written by Himself: a New Critical Edition (CA: City Lights Books, 2010).

the cost of the farm or the land itself. Enslavers mortgaged their human labor enabling companies to enrich themselves while staying away from the cruel business of slavery themselves. Investors in Europe packaged loans based on enslaved bodies while they could publically condemn slavery in their parlors and offices. While the slave trade was abolished in 1806 in Great Britain, the wealth from credit based in enslaved bodies continued to expand and accumulate into the hands of the monetary elite.

Slave-backed mortgages became more and more popular in the North and in Europe well through the Civil War, while slavery itself was seen by most non-slave-owners as repugnant.

The wealth is not to be underestimated. Mathew Desmond says, "so at the height of slavery, the combined value of enslaved workers exceeded that of all the railroads and all the factories in the nation."[3] This statistic is so staggering to any student of history that this interlocutor, Hannah-Jones, pauses in this moment of the podcast and has Desmond repeat it.

No societal arrangement could be—or ever was—more evil than slavery. No institution is more antithetical to the founding values of America, famously recited by school children today. Yet slavery comprehensively dominated the United States from 1619 to 1865. Slavery was not parenthetical to the expansion of America; chattel slavery was the engine of expansion and wealth accumulation in both the North and the South. The invention of race provided a way for white elites from Georgia to Massachusetts to exclude 2 percent of the population from the land and the profits it would provide and instead profit from their labor. It simultaneously justified the concentration of wealth into the hands of a few wealthy elites since white laborers, through the machinery of racism, could identify with the elites without ever being able to attain their financial status. As Nikole Hannah-Jones writes in the 1619 Project, "our democracy's founding ideals were false when they were written. Black Americans have fought to make them true."[4]

Neither slavery nor capitalism is mentioned in the Constitution and yet it is upon this machinery of human private property that America grew. Slavery was the systematic and barbaric extraction of labor from human beings without recompense. It depended upon an extraordinary amount of violence. Its physical

---

3 Mathew Desmond, "The Economy That Slavery Built," 1619 Project: https://www.nytimes.com/2019/08/30/podcasts/1619-slavery-cotton-capitalism.html.

4 Nikole Hannah-Jones, "Essay," 1619 Project. https://www.nytimes.com/interactive/2019/08/14/magazine/1619-america-slavery.html.

brutality made it odious to many white Americans throughout its history. Yet it is in the financial trading upon such bodies that allows European and American elites to both remove themselves from the morality of slavery and intensively profit from it. The habit of extraction, as well as the financial investment of such extraction, was born in slavery. Yet, even four hundred years after the first Africans arrived in America, this habit of trading extracted resources has only intensified in the trade of oil.

American chattel slavery began with a pirate ship bringing twenty to thirty men of the Ndongo and Kongo tribes to Jamestown, Virginia. Originally kidnapped by the Portuguese colonial government, several hundred men had been forced onto a slave ship headed to Veracruz of New Spain. While many dozens died aboard the ship, their voyage was further complicated when two pirate ships captured the ship. One of these, the *White Lion*, docked at Jamestown and sold the men, as indentured servants, for food and supplies to carry on their voyage.

White supremacy was born in the decades that followed as slavery became a race caste system which would dominate American life. In the early 1600s, white and black laborers shared class solidarity. As indentured servants, both white and black, ended the contracts of their labor, they began to demand their freedom and land titles that they were owed by their contracts in order to begin a life in the New World. In the uprising known as Bacon's rebellion, both free blacks and whites demanded better economic conditions from white landowners. To put this uprising down, the colonial government invented whiteness as a category for legal rights, that alongside the category of man, functioned to exclude all blacks as well as women from the rights white men enjoyed.

The next half century saw the passage of hundreds of laws that guaranteed the rights of habeas corpus and property ownership to white men. Meanwhile, black men were subjected to the strict enforcement of anti-miscegenation laws,[5] faced punishment for often trumped up crimes without recourse to a legal system, and denied land titles even after serving indentured contracts. As the cotton gin lubricated the machinery of capitalism in both the fields of the South and the factories of the North, a racial caste system developed that made blackness a lifetime sentence. This caste system girded segregation throughout slavery, reconstruction, Jim Crow, and the mass incarceration system today.

---

5 Jacqueline Battalora, *Birth of a White Nation: The Invention of White People and Its Relevance Today* (Houston: Strategic Publishing, 2013).

Even after the abolition of the slave trade in Great Britain, London investors bought and traded packaged loans and credit systems of black bodies mortgaged by banks throughout the United States. In other words, slavery was an inextricable foundation for the beginning of capitalism. A second key point is that the US South was based on a caste system—whites/owners, black/powerless slaves, and landless/powerless whites. Movement from one caste to another was virtually impossible. Only recently have landless/powerless whites received attention from scholars, and we will highlight their significance below.

Capitalism as it brutalizes working Americans today was created with slavery and the two went hand in hand in the United States. First advanced by Eric Williams in 1944,[6] the thesis has been expanded and substantiated with new evidence, including in a 2019 *New York Times* series. The *New York Times* presentation begins with this statement: "If you want to understand the brutality of American capitalism, you have to start on the plantation."[7] The sheer scale of that brutality must be underscored. In the years between 1790 and 1860 there typically were about twice the number of free whites as enslaved blacks. For example, in 1790 there were just over 654,000 enslaved blacks and 1,240,454 free whites.[8]

While slavery birthed capitalism, it is specifically a brutal, but also, law-aversive type of capitalism that we see today in white-collar crime and manipulation of the markets. White supremacy has allowed the freedom of the market to consistently outpace reconciling itself with the promises of the Constitution allowing wealth to accumulate into the hands of fewer and fewer rich, mostly white and male, since the American Revolution. The same goes for the profit motive's relentless war on the environment such that the environmental justice movement is always biting at the heels of environmental criminals after the damage has been done.

The wages, revenue, land, and property equity that were stolen by the earliest white capitalists on the fields of Georgia and in the mills of Lowell

---

6 Eric Williams, *Capitalism and Slavery* (University of North Carolina Press, 1994); Greg Grandin, *The Empire of Necessity: Slavery, Freedom, and Deception in the New World* (New York: Macmillan, 2014); Sven Beckert, *Empire of Cotton: A Global History* (Penguin Random House, 2015).

7 Desmond, Mathew. "Episode 2: The Economy That Slavery Built," the *New York Times*, 31 August 2019, www.nytimes.com/2019/08/30/podcasts/1619-slavery-cotton-capitalism.html.

8 Economic History Association. "Slavery in the United States": https://eh.net/encyclopedia/slavery-in-the-united-states/.

were never recompensated to their rightful owners.[9] In the newly published seminal book on reparations, *From Here to Equality: Reparations for Black Americans in the Twenty-First Century*, William Darity and Kirsten Mullen catalogue the systematic, three centuries-long theft of black wealth by whites from slavery, through Jim Crow and the new Jim Crow, systemic discrimination today. After the Civil War the federal government failed to endow "forty acres and a mule" to formerly enslaved men, as had been repeatedly been promised. In the wake of that initial failure, dozens of other plans to repair the unspeakable moral and economic costs of slavery were paraded before the government, all to be lost to the dustbins of history. At some point, black abolitionists saw an all too familiar pattern.[10] The case for reparations today is based on the accumulated wealth extracted since the beginning of slavery on these shores in 1619, but also, existentially, on the consistent failure of white society to confront and account for its sins.

This from a recent article published in the *Chronicle of Higher Education:*

We cannot know if the cotton industry was the only possible way into the modern industrial world, but we do know that it was the path to global capitalism. We do not know if Europe and North America could have grown rich without slavery, but we do know that industrial capitalism and the Great Divergence [varied regional development in the United States] in fact emerged from the violent caldron of slavery, colonialism, and the expropriation of land. In the first 300 years of the expansion of capitalism, particularly the moment after 1780 when it entered into its decisive industrial phase, it was not the small farmers of the rough New England countryside who established the United States' economic position. It was the backbreaking labor of unremunerated American slaves in places like South Carolina, Mississippi, and Alabama.[11]

---

9 See Greg Grandin, "Capitalism and Slavery: Each Generation Seems Condemned to Have to Prove the Obvious Anew: Slavery Created the Modern World, and the Modern World's Divisions Are the Product of Slavery." The *Nation*, May 1, 2015: https://www.thenation.com/article/capitalism-and-slavery/.

10 William A. Darity Jr. and A. Kirsten Mullen, *From Here to Equality: Reparations for Black Americans in the Twenty-First Century* (Chapel Hill, NC: University of North Carolina Press, 2020).

11 Sven Beckert, "Slavery and Capitalism," *Chronicle of Higher Education*, December 12, 2014: https://www.chronicle.com/article/SlaveryCapitalism/150787.

Historian Julia Ott writes that, according to Marx, "Global commerce in slaves and the commodities they produced gave rise to modern finance, to new industries, and to wage-labor in the eighteenth century."[12] In other words, slavery laid the foundations of capitalism. Karl Marx believed slavery was despicable and must be ended, which is evident in a co-authored letter sent to Abraham Lincoln in 1864:

Sir:

We congratulate the American people upon your re-election by a large majority. If resistance to the Slave Power was the reserved watchword of your first election, the triumphant war cry of your re-election is *Death to Slavery*.

From the commencement of the titanic American strife the workingmen of Europe felt instinctively that the star-spangled banner carried the destiny of their class. The contest for the territories which opened the dire epopee, was it not to decide whether the virgin soil of immense tracts should be wedded to the labor of the emigrant or prostituted by the tramp of the slave driver?

When an oligarchy of 300,000 slaveholders dared to inscribe, for the first time in the annals of the world, "slavery" on the banner of Armed Revolt, when on the very spots where hardly a century ago the idea of one great Democratic Republic had first sprung up, whence the first Declaration of the Rights of Man was issued, and the first impulse given to the European revolution of the eighteenth century; when on those very spots counterrevolution, with systematic thoroughness, gloried in rescinding "the ideas entertained at the time of the formation of the old constitution", and maintained slavery to be "a beneficent institution", indeed, the old solution of the great problem of "the relation of capital to labor", and cynically proclaimed property in man "the cornerstone of the new edifice" ....

While the workingmen, the true political powers of the North, allowed slavery *to defile their own republic*, while before the Negro, mastered and sold without his concurrence, they boasted it the highest prerogative of the white-skinned laborer to sell himself and choose his own master, they were unable to attain the true freedom of labor, or to support their European brethren in

---

12 Julia Ott, "Slaves: The Capital That Made capitalism," Public Seminar. April 9, 2014. http://www.publicseminar.org/2014/04/slavery-the-capital-that-made-capitalism/.

their struggle for emancipation; but this barrier to progress has been swept off by the red sea of Civil War.

The workingmen of Europe feel sure that, as the American War of Independence initiated a new era of ascendancy for the middle class, so the American Antislavery War will do for the working classes. They consider it an earnest of the epoch to come that it fell to the lot of Abraham Lincoln, the single-minded son of the working class, to lead his country through the matchless struggle for the *rescue of an enchained race* and the reconstruction of a social world.

Signed on behalf of the International Workingmen's Association, the Central Council.[13]

————

Lincoln delivered the Gettysburg address in 1863—in which he proclaimed, "all men are created equal."[14] The Civil War ended in 1865, and later, beginning in 1916, blacks fled the South in large numbers to the North to search for employment, in what is called the Great Migration. Yet very large numbers of immigrants from Europe arrived in the United States in the same years, and with white supremacy still institutionalized in the North blacks were disadvantaged. About 20 million European whites arrived between 1880 and 1920 while around 6 million American blacks migrated North, experiencing vicious discrimination. Until this day, blacks experience discrimination throughout the United States—north, south, east, and west—and this discrimination is comprehensive, including in education, housing, banking services, medical services, and employment.[15]

TODAY'S AMERICA

Usually white Americans experience increases in incomes over, say, a decade, but this has not been true for white or black Americans over the last decades.

————

13 International Workingmen's Association, Address to Abraham Lincoln (Written by Marx): https://www.marxists.org/archive/marx/iwma/documents/1864/lincoln-letter.htm (emphasis added).

14 Abraham Lincoln. Gettysburg Address, November 19, 1863. http://www.abrahamlincolnonline.org/lincoln/speeches/gettysburg.htm.

15 This is widely documented, for example, Marianne Bertrand and Sendhil Mullainathan, "Discrimination in the Job Market in the United States," Poverty Action Lab: https://www.povertyactionlab.org/evaluation/discrimination-job-market-united-states.

In the United States, prime-age workers in the bottom 60 percent have had no real (i.e., inflation-adjusted) income growth since 1980. Yet this has been the time when incomes for the top 10 percent have doubled and those of the top 1 percent have tripled.[16] Moreover in the United States the top 1 percent have seen nearly a 600 percent increase since 1989. The percentage of children who grow up to earn more than their parents has fallen from 90 percent in 1970 to 50 percent today.[17] To be sure the United States is no longer a caste society, as it was in the South before 1865. But there is a lower class and descendants of slaves are so marginalized that it is as if they made up the lowest caste.[18]

GLOBAL INEQUALITY AND POVERTY

According to the UN, more than 700 million people, or 10 percent of the world's population, still live in extreme poverty and are struggling to fulfill the most basic needs like health, education, and access to water and sanitation. Worldwide, the poverty rate in rural areas is 17.2 percent—more than three times higher than in urban areas. Impoverishment can lead to violence, participation in extreme and fringe political and social movements, and rebellion. Oxfam International has been tracking global inequality and concludes:

> Shocking news: Just 26 people now have the same wealth as the poorest 3.8 billion people. Far from trickling down, income and wealth are instead being sucked upwards at an alarming rate.
>
> We need a political system that is accountable to all of us, not just special interests and the wealthy. *It's time to demand a change.*

16 Ray Dalio, "Why and How Capitalism Needs to Be Reformed." Economic Principles, April 5, 2019: https://economicprinciples.org/Why-and-How-Capitalism-Needs-To-Be-Reformed/.

17 Richard V. Reeves and Katherine Guvot, "Fewer Americans...." July 25, 2018:https://www.brookings.edu/blog/up-front/2018/07/25/fewer-americans-are-making-more-than-their-parents-did-especially-if-they-grew-up-in-the-middle-class/; Federal Reserve, "Distributional Financial Accounts: Levels of Wealth by Wealth Percentile Groups": https://federalreserve.gov/efa/efa-distributional-financial-accounts.htm; Drew DeSilver, "For Most U.S. workers, Real Wages Have Barely Budged in Decades,", Pew Research Center, August 7, 2018: https://www.pewresearch.org/fact-tank/2018/08/07/for-most-us-workers-real-wages-have-barely-budged-for-decades/.

18 Angela Hanks, Danyelle Solomon, and Christian E. Weller, "Systematic Inequality." Center for American Progress, February 21, 2018: https://www.americanprogress.org/issues/race/reports/2018/02/21/447051/systematic-inequality/.

Everyone should be able to earn enough to provide for their family, save for the future, and have a fair chance to get ahead. But many hardworking people struggle just to make ends meet, and don't have opportunities to succeed.

Extreme poverty and inequality are the result of a skewed economic and political system that favors the few at the expense of everyone else. Decades of government deregulation and bad political choices have allowed big corporations and billionaires to use their immense influence on political leaders to drown out the voices of hardworking people and rig the rules to their advantage. The people who make our economy run are falling farther and farther behind, so the engine of economic growth breaks down.

We need political and economic reform to level the playing field. Governments, businesses, and citizens around the world must come together and pursue a set of bold actions that will give us a system that creates opportunity for everyone, not just a few.

The French Inequality Lab also devotes effort to tracking global income and wealth. It highlights the following:

At the global level, the top 1% of the income distribution captured 27% of total growth—that is, twice as much as the share of growth captured by the bottom 50%. The top 0.1% captured about as much growth as the bottom half of the world population. Therefore, the income growth captured by very top global earners since 1980 was very large, even if demographically they are a very small group.

Evidence points towards a rise in global wealth inequality over the past decades. At the global level—represented by China, Europe, and the United States—the top 1% share of wealth increased from 28% in 1980 to 33% today, while the bottom 75% share hovered around 10%. Inequality within African countries is increasing.[19]

WHAT DOES SLAVERY HAVE TO DO WITH THIS? PLENTY

To affirm what Americans already know, Glenn C. Loury points out:

Nevertheless, as anyone even vaguely aware of the social conditions in contemporary America knows, we still face a "problem of the color line."

---

19 World Inequality Database: https://wid.world/wid-world/.

The dream that race might someday become an insignificant category in our civic life now seems naively utopian. In cities across the country, and in rural areas of the Old South, the situation of the black underclass and, increasingly, of the black lower working classes is bad and getting worse. No well-informed person denies this, though there is debate over what can and should be done about it. Nor do serious people deny that the crime, drug addiction, family breakdown, unemployment, poor school performance, welfare dependency, and general decay in these communities constitute a blight on our society virtually unrivaled in scale and severity by anything to be found elsewhere in the industrial West.[20]

Research by Grazilla Bertocci and Arcangelo Dimico of Queen's University find that those US counties that had large slave populations have the largest inequality today, and they suggest that the cause is the political exclusion of former slaves and the resulting negative influence on the local provision of education.[21] The conclusion is that US counties that in the past exhibited a higher slave share of the total population turn out to be still more unequal in the present day.[22] And African countries that were the source of slaves have the highest unemployment of African countries today: Angola (29 percent), Congo (46 percent), Mali (8 percent), Senegal (16 percent), Guinea-Bissau, (6 percent) Republic of Congo (11 percent), and Nigeria (23 percent).[23]

CONCLUSIONS

A legacy of slavery for the African continent is that the sources of slaves are today countries with the highest levels of inequality and unemployment. A legacy of slavery for the United States is that the counties that most relied

---

20 Glenn C. Loury, "An American Tragedy," *Brookings*, March 1, 1998: https://www.brookings.edu/articles/an-american-tragedy-the-legacy-of-slavery-lingers-in-our-cities-ghettos/.

21 Graziella Bertocchi and Arcangelo Dimico, "Slavery, Education, and Inequality." *European Economic Review* 70: 197–209: https://www.sciencedirect.com/science/article/pii/S0014292114000695.

22 Tom Jacobs. "Slavery's Legacy: Race-Based Economic Inequality." *Pacific Standard*, June 14, 2017: https://psmag.com/economics/slaverys-legacy-race-based-economic-inequality-83854.

23 Unemployment Rate. *Trading Economics:* https://tradingeconomics.com/country-list/unemployment-rate?continent=africa.

on slavery have the greatest racial inequality today. And African countries that were the source of slaves have the highest unemployment today. Slave states in 1857 included the following: Alabama, Arkansas, Delaware, Florida, Georgia, Kentucky, Louisiana, Maryland, Mississippi, Missouri, North Carolina, South Carolina, Tennessee, Texas, and Virginia. After 1861, these were the slave states: Alabama, Arkansas, Florida, Georgia, Louisiana, Mississippi, North Carolina, South Carolina, Tennessee, Texas, and Virginia. The following slave states remained in the Union: Delaware, Kentucky, Maryland, and Missouri.[24] In other words, North versus South, slave versus non-slave was relatively fluid.

The connection between slave control and management, on the one hand, and capitalism, on the other, has been comprehensively developed and documented, building on the evidence presented first by Eric Williams in his pioneering 1944 book, *Capitalism and Slavery,* and now that connection is broadly accepted by historians. Most recently and forcefully is the assertion made by the *New York Times'* 1619 Project that slave owners would distance themselves from the cruelty of their barbarism by routinizing and standardizing their treatment of slaves. Capitalism was born.

In the account offered here it is hypothesized that in southern states, poor free whites in competition with black slaves not only helped to fuel the Civil War but reinforced the methods and the logic of emerging capitalism.

24 Jonathan R. Allen, "Free and Slave States Map – State, Territory, and City Populations. The Civil War": http://www.nellaware.com/blog/free-and-slave-states-map-state-territory-and-city-populations.html; "Slave States," *World Population Review*: http://worldpopulationreview.com/states/slave-states/.

CHAPTER 3

ARE WE HELPLESS? OR EMPOWERED?

There are three—yes three—horrific global concerns. Each threatens the planet and the well-being of humanity. First, attention now centers on COVID-19; this is totally understandable since it is global, strikes virtually randomly, and is often deadly. It is now identified as a global pandemic, and indeed, a very recent one; it was first reported to the World Health Organization (WHO) on December 31, 2019. The response has been extraordinary. There are now indications of global cooperation. Not all countries are officially cooperating but when their leadership fails or sends out insufficient messages, their people turn to one another or to the international community. The UN WHO agency has been extraordinary.

There are two other global concerns. One is global inequality. As Oxfam reported at 2019 World Economic Forum, the richest 1 percent in the world have more than double the wealth of 6.9 billion people.[1] The second is that race is a good predictor of personal wealth, especially in the United States with its history of slavery. Key findings of a detailed study of racial differences carried out by the Institute of Policy Studies are the following:

- Between 1983 and 2016, the median Black family saw their wealth drop by more than half after inflation, compared to a 33% increase for the median White household. Meanwhile, the number of households with $10 million or more skyrocketed by 856%
- The median Black family today owns $3,600—just 2% of the wealth of the median White family. The median Latino family owns $6,600—just 4% of the median White family.

---

1 "Oxfam Makes Annual Reports Including Ones Presented at the World Economic Forum": https://www.weforum.org/agenda/2020/01/5-shocking-facts-about-inequality-according-to-oxfam-s-latest-report/.

- At this rate, by 2050, median White wealth will be $174,000, while Latino wealth will be $8,600 and Black median wealth will be $600. Black family wealth is on track to reach zero wealth by 2082.[2]

Yes, this is racism. And this is also the case currently for the entire world. In fact, historically and currently white supremacy explains and clarifies how the predominantly white Global North exploits the predominantly non-white Global South. While the collective desire to get rich through the accumulation of capital may have served as a primary motive force for the diffusion of capitalism all over the globe, its movement beyond the United States and Europe relied on the accumulation of racial power in the form of a diffusion of white supremacy. Two determinations operate in the present: (1) white supremacy as a material system of relations, production, ideologies, cultures, and social policies, which works to create and safeguard (2) capitalist-class processes ensure the accumulation of power for white supremacy.

What drove industrialization was initially the ownership and exploitation of black slaves, and as that yielded high returns this in turn drove imperialism and colonialization. There was no concern for the environment; no concern for local culture; no concern for the workers; no concern for nature. Now we bear the consequences. The Intergovernmental Panel on Climate Change (IPCC) has spelled out the disastrous environmental and humanitarian consequences should the Earth warm more than 1.5 degrees Celsius over pre-industrial levels.[3] To prevent this from happening—which we must do—carbon emissions must be slashed to net zero by around 2050. The IPCC report lays out a series of scenarios in which the world is kept from warming over the 1.5°C threshold. With few exceptions, countries are far from meeting targets, and, in fact 2010–19 was the hottest on record with no slowdown in CO_2 emissions.[4] Without question renewables must make up 70 to 85 percent of electricity by 2050.

---

2 Institute for Policy Studies. "Dreams Deferred"; https://ips-dc.org/racial-wealth-divide-2019/.

3 Intergovernmental Panel on Climate Change. "Global Warming of 1.5 °C": https://www.ipcc.ch/sr15/.

4 World Meteorological Organization (WMO). "United in Science." 2019: https://public.wmo.int/en/media/press-release/landmark-united-science-report-informs-climate-action-summit.

DISCOVERY AND SLOW RESPONSE

Let us briefly retrace the history. In 1938 Stewart Callendar linked global warming to CO_2 emissions, but it took over a century for his discovery and message for scientists and the public to understand the relevance of that discovery for planetary change.[5] It wasn't until 1988 that scientists received support to launch international meetings.

In 1988 the IPCC was established through a cooperative agreement by the United Nations Environment Programme (UNEP) and the World Meteorological Organization (WMO). The IPCC is an organization of governments that are members of the United Nations or WMO. The IPCC currently has 195 members. Thousands of people from all over the world contribute to the work of the IPCC. IPCC scientists volunteer their time to assess the thousands of scientific observations, since the objective of the IPCC is to provide governments at all levels with scientific information that they can use to develop climate policies. IPCC reports are also a key input into international climate change negotiations. For the assessment reports, IPCC volunteer-scientists assess the thousands of scientific papers published each year to provide a comprehensive summary of what is known about climate change, its impacts and future risks, and how adaptation and mitigation can reduce those risks.

The United Nations Framework Convention on Climate Change (UNFCCC) is an international treaty adopted on May 9, 1992, and opened for signature at the Earth Summit in Rio de Janeiro from June 3–14, 1992. It then entered into force on March 21, 1994, after a sufficient number of countries had ratified it. The UNFCCC's objective is to "stabilize greenhouse gas concentrations in the atmosphere at a level that would prevent dangerous anthropogenic interference (that is, caused by humans) with the climate system."[6] It is the UNFCCC that hosts the popular and informative web page (https://unfccc.int/), which posts information about meetings and projects, and provides scientific assessments about climate change. It also prepares comprehensive reports under the treaty. Industrialized countries are expected to be in the forefront of reducing emissions and to contribute to the Green Climate Fund that provides financial support to developing countries to mitigate the impact of climate change.

---

5 Spencer R. Weart, *Discovery of Global Warming* (Cambridge: Harvard University Press, 2008).
6 United Nations Climate Change. "Global Climate Action": https://unfccc.int/climate-action.

IN A NUTSHELL

People are discovering that the planet is heating at a faster rate than predicted by scientists. The top part of the ocean is warming up about 24 percent faster than it did a few decades ago, and that rate is likely to increase in the future.[7] Previous work had focused on estimating the damage if average temperatures were to rise by a larger number, 3.6 degrees Fahrenheit (2 degrees Celsius), because that was the threshold scientists previously considered for the most severe effects of climate change. However, if greenhouse gas emissions continue at the current rate, the atmosphere will warm up by as much as 2.7 degrees Fahrenheit (1.5 degrees Celsius) above pre-industrial levels by 2040, inundating coastlines and intensifying droughts and poverty.[8] Besides, storms are increasing in frequency and intensity, and near the equator it is so hot that farming is out of the question. A central concern is that Trump pulled the United States out of the Paris Agreement, which means that the peoples and enterprises of the world's biggest economy are not responding rationally or adequately to urgent climate developments. Global commitment and international support for the goal to slow planetary heating is essential and when a country the size and wealth of the United States does not participate in this process, it is disastrous for the entire world.[9] Besides that, America's nonparticipation has meant that many Americans have become skeptical about the veracity of science, and even now, changing weather patterns and an increase in storms. A remarkably high percentage of Americans are doubtful about climatic warming; according to a recent international YouGov poll, only 36 percent of Americans are certain that humans are chiefly responsible for climate warming. In contrast, 71 percent of people polled in India recognize that the Earth is heating up and humans are responsible.[10]

---

7 Alejandra Borunda, "Ocean Warming Explained," National Geographic: https://www.nationalgeographic.com/environment/oceans/critical-issues-sea-temperature-rise/.

8 Ibid.

9 Judith Blau, *Crimes against Humanity: Climate Change and Trump's Legacy of Planetary Destruction* (New York: Routledge, 2019).

10 Eric Worall, "Yougov Poll: Only 36% of USA Believe Humans Are Mainly Responsible for Climate Change," *Watts Up with That?*: https://wattsupwiththat.com/2019/09/18/yougov-poll-only-36-of-usa-believe-humans-are-mainly-responsible-for-climate-change/.

The IPCC concludes that emissions from fossil fuels need to be cut by half by 2030 and be completely eliminated by 2050 and replaced by renewable energy sources. Even those actions would still leave the world with only 50-50 chance of staying below 1.5 degrees. Scientists and advocates are calling for far faster action to address the climate emergency.[11]

In the next section of this chapter we highlight recent international reports that focus on climate and the warming of the planet. We also ask whether or not the public —especially in the United States—are engaged with science, and whether or not scientists are failing to translate their conclusions in ways that engage non-scientists. Actually, public in most parts of the world are aware that the Earth is getting warmer.

OCEAN

The Special Report on Ocean and Cryosphere in a Changing Climate (SROCC) was approved on September 24, 2019, by 195 UN member governments, and provides new evidence of why it is imperative to reduce global warming and to do so quickly.[12] (See Box 3.1.) The Special Report had been prepared by 104 scientists from 36 countries, and the Summary for Policymakers was presented at a press conference on September 25, 2019. The term "ocean" is clear enough and the term "cryosphere" refers to portions of the Earth's surface where water is in solid form, including sea ice, lake ice, river ice, snow cover, glaciers, ice caps, ice sheets, and frozen ground.

The global ocean covers 71 percent of the Earth surface and contains about 97 percent of the Earth's water. Around 10 percent of Earth's land area is cryosphere that is covered by glaciers or ice sheets. The ocean and cryosphere support unique habitats and are interconnected with other components of the climate system through global exchange of water, energy, and carbon. Without the world's oceans, climate change would be much worse.

---

11 Intergovernmental Panel on Climate Change (IPCC). Press Release. "Choices Made Now Are Critical for the Future of Our Ocean and Cryosphere": https://www.ipcc.ch/site/assets/uploads/2019/09/srocc-P51-press-release.pdf.

12 Intergovernmental Panel on Climate Change. "The Ocean and Cryosphere in a Changing Climate," 2019: https://report.ipcc.ch/srocc/pdf/SROCC_FinalDraft_FullReport.pdf.

BOX 3.1 SPECIAL REPORT OF THE OCEAN AND CRYOSPHERE IN A CHANGING CLIMATE (SROCC). KEY POINTS[a]

- If emissions continue to increase, global sea levels could rise by more than three feet by the end of this century—about 12 percent higher than estimated as recently as 2013.
- To date, the ocean has taken up more than 90 percent of the excess heat in the climate system. By 2100, the ocean will take up 2 to 4 times more heat than between 1970 and the present if global warming is limited to 2°C, and up to five to seven times more at higher emissions. Ocean warming reduces mixing between water layers and reduces the supply of oxygen and nutrients for marine life.
- For decades, the oceans have served as a crucial buffer against global warming, soaking up roughly a quarter of the carbon dioxide that humans emit from power plants, factories, and cars, and absorbing more than 90 percent of the excess heat trapped on Earth by carbon dioxide and other greenhouse gases. Without that protection, the land would be heating much more rapidly.
- The oceans themselves are becoming hotter, more acidic, and less rich in oxygen. If humans keep pumping greenhouse gases into the atmosphere at the current rate, marine ecosystems already facing threats from seaborne plastic waste, unsustainable fishing practices, and other man-made stresses will be further strained.
- Oceans directly absorb about a quarter of the CO_2 spewed into the atmosphere, and they take most of the heat generated by global warming, and they are a buffer against even greater warming. But though they protect us, the oceans also are in great distress.
- Acidic oceans are more hostile to corals, and shellfish, as this decreases the concentration of carbonate, a key ingredient to building shells.
- Glaciers could lose a fifth of their mass this century even if emissions are low.
- Arctic sea ice has declined in all months of the year and around half the summer loss is due to human-caused warming.
- Greenland melt is unprecedented in at least 350 years. With rising Antarctic loss, ice sheets are now contributing 700 percent more to sea levels than two decades ago.

- Arctic near-surface permafrost faces "widespread disappearance," with a 30–99 percent decrease in area if emissions are very high, releasing billions of tons of CO_2.
- The rate of sea-level rise is accelerating and is "unprecedented" over the past century. Worst-case projections are higher than thought and a 2-meter rise by 2100 is not impossible.
- Warming could drastically alter migration flows. If emissions are high, some island nations are likely to become uninhabitable.
- Marine mammals could decline by 15 percent and fisheries by 25 percent this century, and almost all coral reefs will disappear even if emissions are low.
- Cyclones, heat waves, and other extremes are becoming more severe and will exceed the limits of adaptation, causing loss and damage.
- Our oceans and the life they sustain are under mounting pressure from multiple threats, including overfishing, climate breakdown, oil drilling, and plastic pollution. Quite simply, they are in crisis.

---

[a] Ibid. See also, Carbonbrief, "In Depth Questions & Answers: The IPCCS Special Report on the Ocean and Cryosphere": https://www.carbonbrief.org/in-depth-qa-the-ipccs-special-report-on-the-ocean-and-cryosphere.

LAND

Land must remain productive to maintain food security as population increases and as climate change has negative impacts on vegetation. This means there are limits to the contribution of land for the cultivation of crops and the planting of seeds and trees. It also takes time for trees and soils to store carbon effectively. Bioenergy needs to be carefully managed to avoid risks to food security, biodiversity, and land degradation. Desirable outcomes will depend on locally appropriate policies and governance systems.

On August 8, 2019, the IPCC released the Special Report on Climate Change and Land (SRCCL).[13] A guiding principle of SRCCL is that while climate change must be addressed there must simultaneously be improvements

---

13 Intergovernmental Panel on Climate Change. "Special Report on Climate Change and Land (SRCCL)": https://www.ipcc.ch/report/srccl/.

in land, food security, nutrition, and ending hunger. It is recognized in SRCCL that climate change is affecting all four pillars of food security: availability (yield and production), access (prices and ability to obtain food), utilization (nutrition and cooking), and stability (disruptions to availability). The Special Report emphasizes that when land is degraded, it becomes less productive, restricting what can be grown and reducing the soil's ability to absorb carbon. This exacerbates climate change, while climate change in turn exacerbates land degradation in many ways. The report notes that food security will increasingly be affected by global warming. This is through declines in food production, increased prices, reduced nutrient quality, and supply chain disruptions. The report points out that there will be more drastic impacts in low-income countries in Africa, Asia, Latin America, and the Caribbean.

The report also highlights that about one-third of food produced is currently lost or wasted. Causes of food loss and waste differ substantially between developed and developing countries, as well as between regions. Reducing this loss and waste would reduce greenhouse gas emissions and improve food security.[14] The report emphasizes that keeping global warming to well below two degree Celsius can be achieved only by reducing greenhouse gas emissions from all sectors including land and food.

BIODIVERSITY

The Convention on Biological Diversity (CBF) came into force on December 29, 1993. It is dedicated to promoting sustainable development, with three main objectives: (1) the conservation of biological diversity (all ecosystems, species, and genetic resources); (2) the sustainable use of the components of biological diversity; and (3) the fair and equitable sharing of the benefits arising out of the utilization of genetic resources, notably those destined for commercial use.[15]

There are 196 state parties to the convention and several advisory bodies have also been established. It stands as a landmark in international law, recognizing for the first time the conservation of biological diversity as an integral part of the development process (on one hand acknowledging that ecosystems, species, and genes must be used for the benefit of humans, but

---

14 Food and Agriculture Organization. "Food Loss and Food Waste": http://www.fao.org/food-loss-and-food-waste/en/.

15 Convention on Biological Diversity. https://www.cbd.int/convention/text/.

simultaneously maintaining that conservation brings significant environmental, economic, and social benefits in return). The United States is the only country that has not ratified the treaty.[16]

The importance of biodiversity cannot be overstated. "Without biodiversity, there is no future for humanity," said Damian Carrington. In a thoughtful essay in the *Guardian,* he noted, "Biodiversity is comprised of several levels, starting with genes, then individual species, then communities of creatures, and finally entire ecosystems, such as forests or coral reefs, where life interplays with the physical environment. These myriad interactions have made the Earth habitable for billions of years. Some examples are obvious: without plants there would be no oxygen and without bees to pollinate there would be no fruits or nuts. Others are less obvious—coral reefs and mangrove swamps provide invaluable protection from cyclones and tsunamis for those living on coasts, while trees can absorb air pollution in urban areas."[17] The website of the UN devoted to climate change devotes considerable coverage to biodiversity, as do the World Wildlife Fund, the Union of Concerned Scientists, and other science organizations.

The United Nations issued its latest report on biodiversity on May 6, 2019. It is called the Intergovernmental Science-Policy Platform on Biodiversity and Ecosystem Services (IPBES). It concluded that nature is declining globally at rates unprecedented in human history—and the rate of species extinctions is accelerating, with grave impacts on people around the world. The summary was approved at the seventh session of the IPBES Plenary meeting (29 April–4 May) in Paris. It was compiled by 145 expert authors from 50 countries working over three years, with inputs from another 310 contributing authors. The report assesses changes over the past five decades, providing a comprehensive picture of the relationship between economic development pathways and their impacts on nature. It also offers a range of possible scenarios for the coming decades. Based on the systematic review of about 15,000 scientific and government sources, the Report also draws (for the first time ever at this scale) on indigenous and local knowledge, particularly addressing issues relevant to Indigenous Peoples and Local Communities.

---

16 Gloria Dickie and Commentary, "The US Is the Only Country That Has Not Signed on to a Key International Agreement to Save the Planet," *Quartz,* December 25, 2016: https://qz.com/872036/the-us-is-the-only-country-that-hasnt-signed-on-to-a-key-international-agreement-to-save-the-planet/.

17 Damian Carrington, "What Is biodiversity and Why Does It Matter to Us?" The *Guardian,* March 12, 2018: https://www.theguardian.com/news/2018/mar/12/what-is-biodiversity-and-why-does-it-matter-to-us.

THE KIDS ARRIVE JUST IN TIME

And There Is Greta!

> Since our leaders are behaving like children, we will have to take the responsibility they should have taken long ago.—Greta Thunberg[18]

Greta has become famous for her candid, critical, and outspoken views about global warming and what she describes as the inadequate response of adults. She spoke at the United Nations Climate Action Summit on September 23, 2019, met with indigenous peoples of the United States, and has been nominated for the Nobel Peace Prize. She shames adults. When she spoke at the UN in September 2019, she confronted her audience by saying: "People are suffering. People are dying. Entire ecosystems are collapsing. We are at the beginning of a mass extinction and all you can talk about is money and fairytales of eternal economic growth." She said, "How dare you!"[19] Yet young people all over the world are tackling the cause of global warming and speaking out in defense of preserving the environment. This not only includes the young people of rich nations but also young people in developing nations.[20]

DEFENDING THE RIGHTS OF THE CHILD

On September 25, 2019, 13 youths—Chiara Sacchi (Argentina); Catarina Lorenzo (Brazil); Iris Duquesne (France); Raina Ivanova (Germany); Ridhima Pandey (India); David Ackley III, Ranton Anjain, and Litokne Kabua (Marshall Islands); Deborah Adegbile (Nigeria); Carlos Manuel (Palau); Ayakha Melithafa (SouthAfrica); Greta Thunberg (Sweden); Raslen Jbelli (Tunisia); and Carl Smith and Alexandria Villaseñor (USA) filed a complaint against Argentina,

---

18 Paul Davies, "25 of Greta Thunberg's Best Quotes," Speech at the 24th Conference of the Parties to the United Nations Framework Convention on Climate Change (UNFCCC): https://www.curious.earth/blog/greta-thunberg-quotes-best-21.

19 Thalif Deen, "A Rising Youth Movement Picks up Where Governments Have Failed." Inter Press News Service, December 21, 2019: http://www.ipsnews.net/2019/09/rising-youth-movement-picks-governments-failed/.

20 Chika Unigwe, "It's Not Just Greta Thunberg: Why Are We Ignoring the Developing World's Inspiring Activists?" The *Guardian,* October 5, 2019: https://www.theguardian.com/commentisfree/2019/oct/05/greta-thunberg-developing-world-activists.

Brazil, France, Germany, and Turkey for violating a Human Rights Treaty, the Rights of the Child.[21]

The case includes the following points. (We have quoted just a few.)

* The climate crisis is a children's rights crisis. Children have an inalienable right to life under the Convention on the Rights of the Child (the "Convention"). The Convention—the most widely ratified human rights instrument in the world—obligates nations to respect, protect, and fulfill children's inalienable right to life, from which all other rights flow. Mitigating climate change is a human-rights imperative.

* Each respondent—Argentina, Brazil, France, Germany, and Turkey—has known about the harmful effects of its internal and cross-border contributions to climate change for decades. In 1992, each signed the United Nations Framework Convention on Climate Change ("Climate Change Convention") and undertook to protect children from the foreseeable threats of climate change. It was clear then that every metric ton of CO_2 that they emitted or permitted was adding to a crisis that transcends all national boundaries and threatens the rights of all children everywhere.

* Finds that by recklessly perpetuating life-threatening climate change, each respondent is violating the petitioners' rights to life, health, and the prioritization of the child's best interests, as well as the cultural rights of the Petitioners from indigenous communities.

* Endorses "Fridays for the Future," foregoing school on Fridays in an effort to bring awareness and spark action to combat climate change.

OUR CHILDREN'S TRUST

The case of *Juliana vs. the United States* is wending its way up to the Supreme Court and involves 21 American youths, and charges the federal government with violating the constitutional rights of youth by perpetuating systems that contribute to climate breakdown. Those young people—who range in age from

---

21 *Sacchi et al. v. Argentina et al.*: http://climatecasechart.com/non-us-case/sacchi-et-al-v-argentina-et-al/;https://internationalclimatelaw.com/blog/2019/09/sacchi-et-al-v-argentina-et-al/; Communication to the Committee on the Rights of the Child: September 23, 2019:https://earthjustice.org/sites/default/files/files/CRC-communication-Sacchi-et-al-v.-Argentina-et-al.pdf.

11 to 23 and hail from all corners of the nation—argue that the constitution gives them and future generations a right to an environment free of climate catastrophe. On the website of Our Children's Trust is the following statement:

> Our Children's Trust elevates the voice of youth to secure the legal right to a stable climate and healthy atmosphere for the benefit of all present and future generations. Through our programs, youth participate in advocacy, public education and civic engagement to ensure the viability of all natural systems in accordance with science. Our mission is to protect earth's atmosphere and natural systems for present and future generations. We lead a game-changing legal campaign seeking systemic, science-based emissions reductions and climate recovery policy at all levels of government. We give young people, those with most at stake in the climate crisis, a voice to favorably impact their futures.

> On March 1, 2019 powerful voices of support for the *Juliana v. United States* youth plaintiffs and their landmark constitutional climate lawsuit filed amicus curiae ("friend of the court") briefs with the Ninth Circuit Court of Appeals. In all, 15 amicus briefs, filed on behalf of a diverse set of supportive communities, including members of U.S. Congress, legal scholars, religious and women's groups, businesses, historians, medical doctors, international lawyers, environmentalists, and more than 32,000 youth under the age of 25, displayed legal support for *Juliana v. United States* to proceed to trial.[22]

IN SUM

One could say that American public have not caught up with the latest knowledge and understanding of climate change.[23] This is a common assessment because the U.S. president withdrew the country from the Paris Agreement. Yes, it is the case that some towns, cities and states have adopted renewable energy policies but the response has not been uniform, and American scientists have not fully engaged research related to climate, which is no doubt due to President Trump's withdrawal from Paris and reductions in the financial support for climate-related

---

22 Our Children's Trust, Mission Statement: https://www.ourchildrenstrust.org/mission-statement.

23 Pew Research: https://www.pewresearch.org/fact-tank/2019/04/18/a-look-at-how-people-around-the-world-view-climate-change/.

research.[24] The United States is not paying the UN its full share of its budget,[25] nor cooperating in the global drive to slow the heating of the planet. The world of course is complex, and some Americans and many American cities, and many, many young people around the globe are now engaged in environmental activism. Many youth, inspired by Greta Thunberg, are devoting Fridays to protesting and advocacy. (Also see Extinction Rebellion.[26]) The aim is to educate and raise consciousness and enact new laws and rules that will slow the warming of the earth.

Westerners' indifference to the contemporary exploitation of peoples in the Global South as well as our indifference to the ruthless manipulation of the natural environment continues to this day. A sad commentary on the United States is its failure to ratify treaties that protect nature and the natural environment. If COVID-19 is related to indifference to human lives and nature—as it surely is—it is now time for a paradigm shift.

---

24 Judith Blau. *Crimes against Humanity: Climate Change and Trump's Legacy of Planetary Destruction* (UK: Routledge, 2018).

25 Zachary R. Dowdy, "US Pays $563 Million, Part of Dues Owed to United Nations." *Newsday,* December 2, 2019: 2https://www.newsday.com/news/world/united-nations-dues-owed-1.39158123.

26 Extinction Rebellion: https://rebellion.earth/.

CHAPTER 4

WHAT REPLACES CAPITALISM? THE CIRCULAR ECONOMY? BLOCKCHAIN?

First of all, it must be stressed, as we have said before, that an exceptionally ugly truth about global capitalism is that the inequality that it has both created and advanced is horrendous. The estimate is that 26 people own the same as 3.8 billion people who make up the poorest half of humanity.[1] Compare that to 66 million children—including 16.2 million American children—who go hungry and do not get adequate food.[2] What are the reasons for this inequality? Are these tiny few—just 26 the strongest in the world? The smartest? The most attractive? The luckiest? The answer to any and to all of these questions is "what nonsense." A very important question is: Is inequality declining? No. Not at all. Capitalism ensures that the rich get richer and the poor get poorer. And there has been a steady rise of within-country inequality.[3]

An additional and serious problem is that there is a limitless and infinite amount of waste generated; products are simply discarded at the end of their cycle.[4] What follows is a description of a model that is based on the idea that when products reach the end of a cycle they are recycled. Not only that, but the process ensures they are maximally processed to further cooperation, inclusiveness,

---

1 Oxfam International. "Five Shocking Facts about Extreme Inequality and How to Even It Up": https://www.oxfam.org/en/even-it/5-shocking-facts-about-extreme-global-inequality-and-how-even-it-davos.

2 Food Aid Foundation, "1 in 7 People Go Hungry": http://www.foodaidfoundation.org/world-hunger-statistics.html; "Feeding America: Hunger in America": https://www.feedingamerica.org/hunger-in-america/child-hunger-facts.

3 Brookings, "Is Inequality Really on the Rise?" https://www.brookings.edu/blog/future-development/2019/05/28/is-inequality-really-on-the-rise/.

4 Paul Mason, *Postcapitalism: A Guide to Our Future* (London: Penguin, 2015).

and participation based on diversity and mutual recognition. In other words, this is different from capitalism. The idea is the "Circular Economy."[5] We will also discuss a new and emerging model—the "Blockchain"—that creates and strengthens inter-community, inter-city, and inter-state communications so there is greater collaboration, trust, and, even, equality.

Remarkably, the World Economic Forum put out a statement about the circular economy on March 25, 2020: "We must start investing in what matters, by laying the foundation for a green, circular economy that is anchored in nature-based solutions and geared toward the public good."[6] This is an amazing statement coming from an international organization that promotes shareholder capitalism, and it is a point that addresses COVID-19.

As concept and practice, the circular economy is also backed by the United Nations because it reduces waste, promotes democracy, and helps to slow global consumerism. It is also backed by various nonprofit entities, governments, and municipalities. It was initially proposed by the Ellen MacArthur Foundation,[7] supported by the United Nations,[8] and adopted by the Finnish government[9] and then by individual US cities, including the city of Phoenix.[10] The Circular Economy is based on the principles of designing out waste and pollution; keeping products and materials in use; and regenerating natural systems. It advances collaboration, peer-to-peer learning, and diverse and inclusive coalitions.[11]

For example, Alaska Airlines converts its unwanted airline seats to handbags and purses; Omni United (a tire company) has partnered with Timberland to

---

5 SITRA. "The Redistribution of Global Wealth Requires a Circular Economy": https:// www.sitra.fi/en/news/redistribution-global-wealth-requires-circular-economy/.

6 World Economic Forum. Could Covid-19 Give Rise to a Greener Global Future? https://www.weforum.org/agenda/2020/03/a-green-reboot-after-the-pandemic/.

7 Ellen MacArthur Foundation: https://www.ellenmacarthurfoundation.org/.

8 Ellen MacArthur Foundation: https://www.ellenmacarthurfoundation.org/news/ the-ellen-macarthur-foundation-signs-new-agreement-with-un-environment.

9 Finland. Ministry of Environment: https://www.ym.fi/en-US/The_environment/ Circular_economy.

10 City of Phoenix. Phoenix Globally Recognized for Circular Economy Efforts. https:// www.phoenix.gov/news/publicworks/1948.

11 United Nations Conference on Trade and Development. "Circular Economy": https:// unctad.org/en/Pages/DITC/Trade-and-Environment/Circular-Economy.aspx; United Nations Department of Economic and Social Affairs. " 'Waste Not, Want Not' - European Union Goes Circular": https://www.un.org/development/desa/en/news/ sustainable/good-practice-circular-economy.html.

design its tires in such a way that allows them to be recycled into Timberland footwear outsoles; Thread collects plastic waste from Haiti and repurposes it into useable fabric; Hewlett Packard recovers plastic from used ink cartridges and toners via the Planet Partners program and recycles them to produce new ink cartridges; Southwest Airlines partners with Loopworks to repurpose its used seat leather into L duffle bags, shoes, soccer balls, thereby saving over 70 percent of CO_2 emissions; and Unilever's program, "Zero-Non-Hazardous-Waste-to-Landfill," saves over 140,000 tons of waste going into landfill. Unilever purchases over 2 million tons of waste packaging every year and recycles the same for fresh packaging. In other words, the Circular Economy is promoted as an innovative solution that will cut greenhouse gases and reduce reliance on steel, plastics, aluminum, and concrete. It could help, according to the UN's climate program—UNFCCC—to achieve net zero emissions by 2050. Moreover, it allows considerable local autonomy, and can be either capitalist or socialist.[12]

To help explain the Circular Economy, we provide two definitions, one by the Ellen-MacArthur Foundation and the other by an international agency—the World Economic Forum (an NGO with over 1,000 multinational members). The Foundation describes circular economy in these terms:

> It replaces the "end-of-life" concept with restoration, shifts towards the use of renewable energy, eliminates the use of toxic chemicals, which impair reuse, and aims for the elimination of waste through the superior design of materials, products, systems, and, within this, business models. Such an economy is based on few simple principles. First, at its core, a circular economy aims to "design out" waste. Waste does not exist—products are designed and optimized for a cycle of disassembly and reuse. [...] Secondly, circularity introduces a strict differentiation between consumable and durable components of a product. Unlike today, consumables in the circular economy are largely made of biological ingredients or "nutrients" that are at least non-toxic and possibly even beneficial and can be safely returned to the biosphere—directly or in a cascade of consecutive uses.[13]

---

12 Jo Confino, "Beyond Capitalism and Socialism." The *Guardian*, April 21, 2015: https://www.theguardian.com/sustainable-business/2015/apr/21/regenerative-economy-holism-economy-climate-change-inequality.

13 Ellen MacArthur Foundation. *Towards the Circular Economy*. 2013: https://www.ellenmacarthurfoundation.org/assets/downloads/publications/Ellen-MacArthur-Foundation-Towards-the-Circular-Economy-vol.1.pdf.

WORLD ECONOMIC FORUM'S SUMMARY:

Linear consumption is reaching its limits. A circular economy has benefits that are operational as well as strategic, on both a micro- and macroeconomic level. This is a trillion-dollar opportunity, with huge potential for innovation, job creation and economic growth.

The last 150 years of industrial evolution have been dominated by a one-way or linear model of production and consumption in which goods are manufactured from raw materials, sold, used and then discarded or incinerated as waste. In the face of sharp volatility increases across the global economy and proliferating signs of resource depletion, the call for a new economic model is getting louder. The quest for a substantial improvement in resource performance across the economy has led businesses to explore ways to reuse products or their components and restore more of their precious material, energy and labor inputs. A circular economy is an industrial system that is restorative or regenerative by intention and design. The economic benefit of transitioning to this new business model is estimated to be worth more than one trillion dollars in savings. [...] The current model is take, make, dispose. There is huge waste of energy, materials, and growing scarcity of natural resources.[14]

To illustrate, within California, farm-to-school programs that support local procurement practices can cultivate the following: (1) economic capital by increasing the net sales, creating new labor income and improving the local purchasing power; (2) natural capital through reduced CO_2 emissions, regenerative practices for land restoration and diversification; and (3) social capital through increased access to healthy food and job creation that may increase local economic activity and improve quality of life. In fact, in 2018, a bill was debated in the California Assembly to advance the practice of Circular Economy, and it was recognized that the practice would reduce plastic pollution.[15]

---

14 World Economic Forum: 1. "The Benefits of the Circular Economy": http://reports. weforum.org/toward-the-circular-economy-accelerating-the-scale-up-across-global-supply-chains/1-the-benefits-of-a-circular-economy/.

15 California. Senate Bill SB 54. "This Bill Would Enact the California Circular Economy and Pollution Reduction Act": https://leginfo.legislature.ca.gov/faces/billTextClient. xhtml?bill_id=201920200SB54.

BLOCKCHAIN TECHNOLOGY

There are various definitions and here we provide the one from Merriam Webster: "a digital database containing information (such as records of financial transactions) that can be simultaneously used and shared within a large decentralized, publicly accessible network also: the technology used to create such a database. The Blockchain is an open, distributed ledger that can record transactions between two parties efficiently and in a verifiable and permanent way."[16] In other words, each segment is a chain of blocks storing digital chunks or pieces of information. Blockchain technology is a relatively new idea, proposed only a couple of years ago; it offers the great advantage that it is democratic, allowing poor with marginalized communities and countries to catch up with their more affluent counterparts.

The UN has described a blockchain as a "dynamic informational database that is updated continuously, democratically, and verifiably by its users, and its users can be people, countries, regions, cities, and businesses." Each added block of data is "chained" and becomes part of a growing list of records, under the surveillance of network members. This technology enables the transfer of assets and the recording of transactions through a secure database. For example, the technology could allow for the development of platforms for peer-to-peer renewable energy trade. Consumers would be able to buy, sell, or exchange renewable energy with each other, using tokens or tradable digital assets representing a certain quantity of energy production.[17] The U.S. Postal Service advocates blockchain because it "offers end-to-end transparency, allowing all parties to view and confirm others' activities. And because the ledger is immutable, meaning it cannot be changed once batches of data are uploaded and verified across the network, tampering or hacking is nearly impossible."[18]

Although there is still a lack of consensus about how blockchains will work, their transparency is attractive as well as offering opportunities for marginalized

---

16 https://www.merriam-webster.com/dictionary/blockchain.
17 United Nations. "Climate Change. How Blockchain Technology Could Boost Climate Action": https://unfccc.int/news/how-blockchain-technology-could-boost-climate-action.
18 USPS. "Is Blockchain Right for Your Supply Chain?": https://www.uspsdelivers.com/is-blockchain-right-for-your-supply-chain/?utm_medium=searchengine&utm_source=google&utm_campaign=cmshipping&utm_content=e516_62cm&gclid=Cj0KCQjwyPbzBRDsARIsAFh15JYUv9S2t9o8FCHQoMC2aZ-11cbVVC8H1Rrm6NU3rhIXiHQpv0BbFY0aAojeEALw_wcB&gclsrc=aw.ds.

communities and countries to catch up. There is great hope that it will support efforts to slow climate change.[19]

IN SUM

The Circular Economy is an economic system that aims to minimize waste and to make the most of resources, and it is promoted by the UN and various world partners in efforts to slow planetary warming. It contrasts with the traditional linear economy, which has a "take, make, dispose" model of production and which does not emphasize reuse. In a circular system input waste, emission, and energy leakage are minimized by slowing, closing, and narrowing energy and material loops, and reusing long-lasting design, as well as emphasizing maintenance, repair, reuse, remanufacturing, refurbishing, and recycling. It should be stressed that a circular economy is an alternative to a traditional linear economy in which we keep resources in use for as long as possible, extract the maximum value from them whilst in use, and then toss and dispose of them. The practices embedded in circular economy greatly reduce waste.

While the circular economy model does not directly address democracy, participation, or decentralization, the blockchain model potentially does because it is transparent and inclusive. The term "blockchain technology" typically refers to the transparent, trustless, publicly accessible ledger that allows people to securely transfer the ownership of units of value using public key encryption and proof-of-work methods. The technology lends itself to decentralization and democracy because it is based on networks that are not controlled by a bank, corporation, or government. In fact, the larger the network grows and becomes increasingly decentralized, the more secure it becomes.

---

19 United Nations. "Climate Change": https://unfccc.int/news/how-blockchain-technology-could-boost-climate-action.

CHAPTER 5

GEOENGINEERING

Many, if not most, scientists assume that the world is on course for a disastrous calamity due to overheating. It is hard to prove this beyond all doubts, but scientists have carried out considerable research that supports this dire prediction, and their efforts have been devoted to understanding and averting this calamity that spells eventual human extinction. If true, there are three different remedies: (1) quickly transitioning to 100 percent renewable energy; (2) prioritizing nature; and (3) adopting geoengineering technologies. Each of these recognizes either implicitly or explicitly that the goal is for CO_2 emissions to reach "net zero" around 2050. What advocates of (1) focus on is ending the use of fossil fuels, while advocates of (2) focus on "protecting nature," and (3) those that advocate major intervention (geoengineering) do so while recognizing that the outcomes are uncertain. In this chapter we highlight geoengineering, and in the next chapter, we review advocacy of "Mother Nature."

The empirical reality is that the Earth is warming faster than predicted, which may play a role in the process of deciding which of the three should be prioritized or adopted. Relevant is the recent study the UN commissioned: "the Special Report on Global Warming of 1.5 degrees Celsius."[1] It is based on the collaboration of 91 climate scientists and the evaluation and feedback of hundreds more. The report unequivocally advocates drastic reduction in fossil fuels to keep emissions of carbon dioxide (CO_2) to about 45 percent lower than 2010 levels by 2030, and to reach zero around 2050.[2] Do note that unless there is rapid de-carbonization there will be high risks of extreme heat, drought, floods,

---

1 Intergovernmental Panel on Climate Change. (IPCC). "Global Warming of 1.5°C": https://www.ipcc.ch/sr15/.
2 United Nations. Secretary-General. "Net Zero Greenhouse Gas Emissions Vital amid Catastrophic Climate Change Scenario," Jan 10, 2019: https://www.un.org/press/en/2019/sgsm19429.doc.htm.

and unprecedented numbers of climate refugees. Already there are places on Earth that are too hot to be habitable.[3] In other words, it may be too late for the rapid transition to renewable energy or the prioritization of the protection of nature. On the other hand, geoengineering shares some of the assumptions that advocates of slavery did—namely, dominance and manipulation.

It is important to stress that the United Nations supports the rapid development and adoption of renewable energy, while the UN's democratic body—the General Assembly—has passed a resolution supporting Mother Earth Day (which is a repudiation of geoengineering that we will discuss in the next chapter). The UN General Assembly has also adopted a resolution, "Harmony with Nature."[4] This resolution affirms that the Earth and its ecosystems are our common home and expresses the conviction that it is necessary to promote Harmony with Nature in order to achieve a just balance among the economic, social, and environmental needs of the present and future generations. Nevertheless, it is important to stress that the global community (specifically, the United Nations) has not ruled out geoengineering but concludes that it is highly risky and must not at this point be the first choice. In 2017 the UN adopted a provisional statement:

> We need to have full-on public engagement, full-on societal involvement. The reason is that the risks of climate change are huge, the risks of doing nothing are huge; but the risks of geoengineering are huge as well. We've got to explore those risks, because who knows, we may end up entering a very risky world without understanding it."[5]

What is geoengineering? Climate engineering or climate intervention, commonly referred to as *geoengineering*, is the deliberate and large-scale intervention in the Earth's climate system, usually with the aim of mitigating the adverse effects of global warming. There are two major approaches: Solar Radiation Management (SRM) and Greenhouse Gas Removal (GGR).

---

3 UN Refugee Agency. "Climate Change and Disaster Displacement": https://www.unhcr. org/en-us/climate-change-and-disasters.html

4 United Nations. Harmony with Nature: http://www.harmonywithnatureun.org/

5 United Nations. Sustainable Development Goals. "Climate Engineering Is Risky, but Should Be Explored Experts Say at UN Conference": https://www.un.org/ sustainabledevelopment/blog/2017/11/climate-engineering-risky-explored-experts-say- un-conference/.

SOLAR RADIATION MANAGEMENT (SRM)

SRM techniques aim to reflect a small proportion of the Sun's energy back into space, counteracting the temperature rise caused by increased levels of greenhouse gases in the atmosphere, which absorb energy and raise temperatures, or that reflect a small fraction of sunlight back into space or increase the amount of solar radiation that escapes back into space to cool the planet. [6] In contrast to carbon geoengineering (see below), SRM does not address the root cause of climate change. It instead aims to break the link from concentrations to temperatures, thereby reducing some climate damages. SRM has the potential to reduce some of the impacts of climate change, but it could also be very risky or rejected.

Introducing aerosols into the stratosphere to absorb energy before it gets absorbed by the earth. There would be constant aircraft in the sky, spraying clouds of aerosols into the stratosphere.

Increasing the water droplet concentration which would brighten marine clouds. This would include spraying fine mists of salt water above marine clouds, which would react with the water droplets to make them smaller, ultimately making the cloud brighter as a whole.

Brightening surfaces of cities, oceans and deserts, eventually to reflect energy off of the earth's surface. All roofs, roads, and pavement would be whitened.

Increasing the reflectance of incoming energy by putting reflective structures into space.

Albedo enhancement. That is, reflecting a small fraction of incoming sunlight back to space in order to attenuate changes in temperature and other climate variables. Increasing the reflectiveness of clouds or the land surface so that more of the Sun's heat is reflected back into space.

---

6 Harvard Solar Engineering Research Program: "Geoengineering":https://geoengineering.environment.harvard.edu/geoengineering; Oxford University. Oxford Geoengineering Program. "What Is Geoengineering": http://www.geoengineering.ox.ac.uk/www.geoengineering.ox.ac.uk/what-is-geoengineering/what-is-geoengineering/indexd41d.html; University of Exeter. "Geoengineering": http://www.exeter.ac.uk/g360/geoengineering/.

GREENHOUSE GAS REMOVAL (GGR) OR CARBON GEOENGINEERING

GGR techniques aim to remove carbon dioxide or other greenhouse gases from the atmosphere, directly countering the increased greenhouse effect and ocean acidification. [7] They would address the root cause of climate change— the accumulation of carbon dioxide in the atmosphere. These techniques would have to be implemented on a global scale to have a significant impact on greenhouse gas levels in the atmosphere. Some proposed techniques include the following:

Afforestation. Engaging in a global-scale tree planting effort. Afforestation is certainly the least invasive, and is, in fact, beneficial since trees have three effects (1) they absorb carbon, pulling from the atmosphere, creating a cooling effect; (2) their dark green leaves absorb light from the sun, heating the earth's surface; and (3) they draw water from the soil, which evaporates into the atmosphere, creating low clouds.

Bio-energy with carbon capture and sequestration. Growing biomass, burning it to create energy and capturing and sequestering the carbon dioxide created in the process.

Ambient Air Capture. Building large machines that can remove carbon dioxide directly from ambient air and store it elsewhere.

Ocean Fertilization. Adding nutrients to the ocean in selected locations to increase primary production which draws down carbon dioxide from the atmosphere.

Enhanced Weathering. Exposing large quantities of minerals that will react with carbon dioxide in the atmosphere and storing the resulting compound in the ocean or soil.

Ocean Alkalinity Enhancement. Grinding up, dispersing, and dissolving rocks such as limestone, silicates, or calcium hydroxide in the ocean to increase its ability to store carbon and directly ameliorate ocean acidification.

The world's leading climate scientists have warned that there is only a dozen years for global warming to be kept to a maximum of 1.5 degrees Celsius,

---

7 The Royal Society. Greenhouse Gas Removal: https://royalsociety.org/topics-policy/projects/greenhouse-gas-removal/.

beyond which even half a degree will significantly worsen the risks of drought, floods, extreme heat, and poverty for hundreds of millions of people. The question is whether or not all countries can agree on the shared strategy to limit warming and to adopt practices that lead to 100 percent adoption of renewable energy: sunlight, wind, rain, tides, waves, and geothermal heat. If this is done rapidly, there would be no need for geoengineering, and to be sure, that depends on the United States to rejoin the rest of the world on the climate treaty. In the next chapter we spell out the implications of the Mother Earth approach.

CONCLUSIONS

Geoengineering is decisive, conclusive, and "top-down" with control exercised by a few over the many. In this sense it shares with capitalism a demagogued approach, and if a geoengineered solution is applied and fails to work, the consequences could be catastrophic. Its defenders often say that sometimes it is the only intervention possible.

CHAPTER 6

HANDS OFF MOTHER EARTH

The Mother Earth Movement is now international, but it initially originated as an indigenous movement, enshrined in the Bolivian Constitution and championed internationally by Bolivian president Evo Morales, Latin America's first indigenous president. Other constitutions that now include this term are those of Ecuador, Mongolia, and Nicaragua. The first paragraph of the Preamble of the 2009 Bolivian Constitution is:

> In ancient times mountains arose, rivers moved, and lakes were formed. Our Amazonia, our swamps, our highlands, and our plains and valleys were covered with greenery and flowers. We populated this sacred *Mother Earth* with different faces, and since that time we have understood the plurality that exists in all things and in our diversity as human beings and cultures. Thus, our peoples were formed, and we never knew racism until we were subjected to it during the terrible times of colonialism.[1]

Allied with an international farm worker movement, via Campesina, the Mother Earth Movement expanded and grew to become the "Hands Off Mother Earth" Movement seeking to discredit geoengineering. Do note its various names and acronyms:

Against Geoengineering (Geoengineering Monitor);[2]
Hands Off Mother Earth (H.O.M.E.);[3]

---

1 Bolivia Constitution: https://www.constituteproject.org/constitution/Bolivia_2009?lang=en. Emphasis added.
2 Geoengineering Monitor. "Against Geoengineering": http://www.geoengineeringmonitor. org/2018/11/against-geoengineering/.
3 Etc Group. "Hands Off Mother Earth": https://www.etcgroup.org/content/hands-mother-earth-0; La Via Campesina "Hands off Mother Earth": https://viacampesina. org/en/hands-off-mother-earth/.

Action Group on Erosion, Technology and Concentration (ETC);[4] Harmony with Nature.[5]

The movement's two goals are to celebrate Nature and to stop all attempts at geoengineering. The history of the movement against geoengineering and how this history intertwines with the Mother Earth movement is documented by Geoengineering Monitor.[6] Probably the most visible around the world is Hands Off Mother Earth (H.O.M.E.) that posts this manifesto:

> We civil society organizations, popular movements, Indigenous Peoples, peasant organizations, academics, intellectuals, writers, workers, artists and other concerned citizens from around the world, oppose geoengineering as a dangerous, unnecessary and unjust proposal to tackle climate change. Geoengineering refers to large-scale technological interventions in the Earth's oceans, soils and atmosphere with the aim of weakening some of the symptoms of climate change. Geoengineering perpetuates the false belief that today's unjust, ecologically- and socially-devastating industrial model of production and consumption cannot be changed and that we therefore need techno-fixes to tame its effects. However, the shifts and transformations we really need to face the climate crisis are fundamentally economic, political, social and cultural.[7]

The H.O.M.E. campaign provides a common platform for organizations and individuals to register their opposition to geoengineering experiments. The campaign asserts that the seas, skies, and soils of our home planet should not be used as a laboratory for these unjust and risky technological fixes. It was launched in April 2010 at The World People's Conference on Climate Change and the Rights of Mother Earth in Cochabamba, Bolivia. It is a coalition of international civil society groups, indigenous peoples' organizations, and social

---

4 Etc Group. "Hands Off Mother Earth": http://www.etcgroup.org/content/hands-mother-earth-0; La Via Campesina "Hands off Mother Earth": https://viacampesina.org/en/hands-off-mother-earth/.

5 United Nations. Harmony with Nature: http://www.harmonywithnatureun.org/.

6 Geoengineering Monitor. "Against Geoengineering": http://www.geoengineeringmonitor.org/2018/11/against-geoengineering/.

7 Geoengineering Monitor. "Hands Off Mother Earth. Manifesto against Geoengineering": http://www.geoengineeringmonitor.org/2018/10/hands-off-mother-earth-manifesto-against-geoengineering/.

movements. These groups invite other organizations worldwide, as well as individuals to sign up in support of the campaign. The campaign proposed that the Universal Declaration of the Rights of Mother Earth be adopted by the UN, or to use the declaration as an inspiration for the drafting of an official UN declaration. It "includes the wisdom and knowledge of people from ancient and modern societies from around the world and points the way to aligning our laws and ways of living with those of Nature."[8] H.O.M.E.— has a website on which supporters can post their support for a UN Declaration of the Rights of Mother Earth.[9] Another group ETC—has goals that are in sync with H.O.M.E., and posted on the ETC website is this statement about geoengineering:[10]

- **It doesn't work:** None of the technologies have a track record, all of them come with major risks and unknowns, and in some cases the effects would be obviously catastrophic.
- **Weaponization:** Computer models show that geoengineering interventions can have regional winners and losers; to the extent that geoengineering successfully changes climate patterns in a predictable way, it will inevitably be weaponized.
- **Detracts from real solutions:** By promising a quick fix, geoengineering threatens to delay the implementation of a transition away from fossil fuels, and could redirect funding and investments away from real climate solutions. Some geoengineering proposals require vast amounts of energy, which means less climate-friendly energy for everyone else.
- **Human rights and biodiversity:** Many geoengineering proposals require the intensive exploitation of vast amounts of land (in the case of BECCS, twice the size of India!). Those projects would inevitably displace millions of people and potentially wipe out entire ecosystems.

It is hard to be opposed to this campaign, and, of course, to protect the Earth from harmful interventions of all kinds, such as the application of harmful

---

8 Universal Declaration of Rights of Mother Earth: https://therightsofnature.org/universal-declaration/.

9 Rights of Mother Earth: https://www.rightsofmotherearth.com/.

10 "Reasons to Oppose Geoengineering." *Geoengineering Monitor.* http://www.geoengineering monitor.org/reasons-to-oppose/.

chemicals. On the other hand, it would be foolish to abandon geoengineering interventions, if doing otherwise would result in certain destruction. Otherwise, it was a major victory for indigenous peoples, via Campesina, and grassroots mobilizers when the UN General Assembly adopted a resolution 63/278 proclaiming that 22 April is International Mother Earth Day. In part it reads as follows:

Resolution Adopted by the General Assembly on 22 April 2009

63/278. INTERNATIONAL MOTHER EARTH DAY

The General Assembly,

Reaffirming Agenda 21,[11] and the Plan of Implementation of the World Summit on Sustainable Development ("Johannesburg Plan of Implementation"),[12]

Recalling the 2005 World Summit Outcome,[13]

Recalling also its resolution 60/192 of 22 December 2005 proclaiming 2008 the International Year of Planet Earth,

Acknowledging that the Earth and its ecosystems are our home, and convinced that in order to achieve a just balance among the economic, social, and environmental needs of present and future generations, it is necessary to promote harmony with nature and the Earth,

Recognizing that Mother Earth is a common expression for the planet earth in a number of countries and regions, which reflects the interdependence that exists among human beings, other living species and the planet we all inhabit,

Noting that Earth Day is observed each year in many countries,

1. *Decides* to designate 22 April as International Mother Earth Day;
2. *Invites* all Member States, the organizations of the United Nations system, international, regional and subregional organizations, civil

---

11 *Report of the United Nations Conference on Environment and Development, Rio de Janeiro, 3–14 June 1992*, vol. I, Resolutions Adopted by the Conference (United Nations publication, Sales No. E.93.I.8 and corrigendum), resolution 1, annex II.

12 *Report of the World Summit on Sustainable Development, Johannesburg, South Africa, 26 August–4 September 2002* (United Nations publication, Sales No. E.03.II.A.1 and corrigendum), chap. I, resolution 2, annex.

13 United Nations. 2005. World Summit. 2005. "World Summit Outcome": http://archive.ipu.org/splz-e/unga05/outcome.pdf. See resolution 60/1.

society, nongovernmental organizations and relevant stakeholders to observe and raise awareness of International Mother Earth Day, as appropriate;

3. *Requests* the Secretary-General to bring the present resolution to the attention of all Member States and organizations of the United Nations system.

80th plenar y meeting
22 April 2009

CHAPTER 7

THE MILLENNIUM DEVELOPMENT GOALS, THE SUSTAINABLE DEVELOPMENT GOALS, AND THE PARIS AGREEMENT

The United Nations was founded in 1945 with the purpose of ensuring global solidarity in the aftermath of WWII. Its initial stated objectives were

To keep peace throughout the world;

To develop friendly relations among nations;

To help nations work together to improve the lives of poor people, to conquer hunger, disease and illiteracy, and to encourage respect for each other's rights and freedoms;

To be a centre for harmonizing the actions of nations to achieve these goals.[1]

These objectives are currently pursued by various entities (what the UN refers to as members of the "family"): six separate funds and programs, including the World Food Programme, UN-Habitat, and the UN Children's Fund or UNICEF; sixteen specialized entities including the Food and Agriculture Organization (FAO), International Labour Organization (ILO), and the World Health Organization (WHO); eight other entities, including the United Nations Framework Convention on Climate Change (UNFCCC). Main bodies include the Security Council, the General Assembly, the High Commissioner for Human Rights, and Human Rights Treaty Bodies, and additional bodies include the Permanent Forum on Indigenous Issues, Special Advisers on the Prevention of

---

1 United Nations, "History": https://www.un.org/en/sections/history/history-united-nations/index.html.

Genocide and the Responsibility to Protect, UN Women, UN Peace Operations, Refugees (UNHCR), and five regional commissions, including the United Nations Economic Commission for Africa (UNECA).[2] The UN's activities are extraordinarily important for promoting peace, development, cooperation, and security and, increasingly, coordinating global action to slow planetary heating. The World Health Organization has been especially important in discovering and promoting a global response—and cure—for COVID-19.

From the beginning, the UN has ambitiously pursued global goals; that is, advancing human welfare while promoting peace and harmony. Some ambitious efforts focus on some countries and not others, whereas others are comprehensive and global. A key example of the former is the Millennial Development Goals (MDGs), 2000–15. And a key example of the latter is the Sustainable Development Goals (SDGs), 2015–30. The MDGs applied only to poor countries, while rich countries supplied the resources. The SDGs apply to all countries. In part, the SDGs deal with climate warming, but it has become clear that the Earth's warming is outpacing efforts to slow it down. Additionally, neoliberalism and corporate practices lead to gross inequalities, impeding development and international cooperation. Blame is appropriately placed (in our view) on multinationals and the very rich who do not have it in their financial self-interest to implement programs that are designed to empower minorities and women, enrich poor countries, and to address climate change.

We start here with the MDGs. To be clear, the MDGs encompassed eight goals that focused entirely on developing countries. Figure 7.1 is a list of these goals, and a brief description of some of the achievements that were made by 2015 as summarized in an UN report.[3]

In the Preface to the 2015 final report, Ban Ki-Moon, Secretary-General of the United Nations, described the progress made toward achieving the Millennium Goals:

> The global mobilization behind the Millennium Development Goals has produced the most successful anti-poverty movement in history. The landmark commitment entered into by world leaders in the year 2000—to "spare no effort to free our fellow men, women and children from the abject

2 UN Organizational Chart. https://www.un.org/en/pdfs/18-00159e_un_system_chart_17x11_4c_en_web.pdf; https://www.orgcharting.com/un-organizational-chart/

3 United Nations, *The Millennium Goals Report* 2015: http://www.un.org/millenniumgoals/2015_MDG_Report/pdf/MDG%202015%20rev%20(July%201).pdf.

Figure 7.1 Millennium Development Goals, 2000–15. Achievements by 2015.

Goal 1 Eradicate extreme poverty and hunger

In 1990 nearly half of the population in the developing world lived on less than $1.25 a day. That dropped to 14 percent by 2015.

Globally, the number of people living in extreme poverty declined by more than half, falling from 19 billion in 1990 to 836 million in 2015.

The proportion of undernourished people in the developing countries fell by almost half, from 23.3 percent in 1990-12.9 percent in 2015.

Goal 2 Achieve universal primary education

The primary school net enrollment rate in the developing regions has reached 91 percent in 2015, up from 83 percent in 2000.

The number of out-of-school of primary school age worldwide fell by almost half, to an estimated 57 million in 2015, down from 100 million in 2000.

The literacy rate among youth aged 15 to 24 has increased globally from 83 percent to 91 percent between 1990 and 2015.

Goal 3 Promote gender equality and empower women

The developing regions as a whole have achieved the target to eliminate gender disparity in primary, secondary and tertiary education.

In Southern Asia, only 74 girls were enrolled in primary school for every 100 boys in 1990. Today 103 girls are enrolled for every 100 boys.

Women now make up 41 percent of paid workers outside the agricultural sector, an increase from 35 percent in 1990.

Goal 4 Reduce child mortality

The global under-five mortality rate has declined by more than half, dropping from 90 to 43 deaths per 1,000 live births between 1990 and 2015.

Since the early 1990's, the rate of reduction of under-five mortality was over five times faster during 2005-2013 than it was during 1990-1995.

About 84 percent of children worldwide received at least one dose of measles-containing vaccine in 2013, up from 73 percent in 2000.

Goal 5 Improve maternal health

Since 1990, the maternal mortality rate has declined by 45 percent worldwide.

More than 71 percent of births were assisted by skilled personnel globally in 2014, an increase from 59 percent in 1990.

Contraceptive prevalence among women, aged 15 to 49, married or in a union, increased from 55 percent worldwide to 64 percent in 2015.

Figure 7.1 (*Continued*)

Goal 6 Combat HIV/AIDS, malaria, and other diseases

By June 2014, 13.6 million people living with HIV were receiving antiretroviral therapy (ART) globally – an immense increase from just 800,000 in 2003. ART averted 7.6 million deaths from AIDS between 1995 and 2013.

Over 6.2 million malaria deaths were averted between 2000 and 2015. The global malaria incidence rate has fallen by an estimated 37 percent and the mortality rate by 58 percent.

Between 2000 and 2013, tuberculosis prevention, diagnosis and treatment interventions saved an estimated 57 million lives.

Goal 7 Ensure environmental sustainability

Ozone-depleting substances have been virtually eliminated since 1990 and the ozone layer is expected to recover by mid-century.

In 2015, 91 percent of the global population is using an improved drinking water source, compared to 76 percent in 1990.

Worldwide, 2.1 billion people have gained access to improved sanitation. The proportion of people practicing open defecation has fallen almost by half since 1990.

Goal 8 Develop a global partnership for development

Official development assistance from developed to developing countries increased by 66 percent from 2000 to 2014 to reach $135.2 billion.

The number of mobile-cellular subscriptions grew almost tenfold in 15 years, from 738 million in 2000 to over 7 billion in 2015.

Internet penetration grew from just over 6 percent of the world's population in 2000 to 43 percent in 2015. As a result, 3.2 billion people are linked to a global network of content and applications.

Source: UN. "Millennium Goals and Beyond, 2015": https://www.un.org/millenniumgoals/bkgd.shtml.

and dehumanizing conditions of extreme poverty"—was translated into an inspiring framework of eight goals and, then, into wide-ranging practical steps that have enabled people across the world to improve their lives and their future prospects. The MDGs helped to lift more than one billion people out of extreme poverty, to make inroads against hunger, to enable more girls to attend school than ever before and to protect our planet. They generated new and innovative partnerships, galvanized public opinion and showed the immense value of setting ambitious goals. By putting people and their immediate needs at the forefront, the MDGs reshaped decision-making in developed and developing countries alike.

Yet for all the remarkable gains, I am keenly aware that inequalities persist and that progress has been uneven. The world's poor remain overwhelmingly concentrated in some parts of the world.[4]

What, of course, made all of this possible is that each of 189 countries volunteered to be a part of this process, whether as a developed or developing country. It was a thoroughly collaborative and cooperative process, setting the precedence for the SDGs and the Paris Agreement. Yet there were no goals in the MDGs aimed at reducing greenhouse gases or combating climate change although throughout this period, from 2000 to 2015, there already were annual international UN-sponsored conferences devoted to climate change. In fact, as earlier noted, there had been world conferences since 1979 devoted to understanding and measuring the extent to which greenhouse gases are warming the Earth.

However, in the final MDG 2015 report attention focused on climate change, providing a launch for the 2016 Sustainable Development Goals (SDGs), which more explicitly address climate change. This 2015 report states:

> Global emissions of carbon dioxide have increased by over 50 per cent since 1990. Addressing the unabated rise in greenhouse gas emissions and the resulting likely impacts of climate change, such as altered ecosystems, weather extremes and risks to society, remains an urgent, critical challenge for the global community.[5]

THE SDGS

The context is extremely important. Keep in mind that the sole focus of the MDGs was on developing countries, but accompanying the expansion of neoliberalism and globalization has been increasing economic inequality; so, while the world proclaimed that rich countries should help poor countries, there are few people who are extremely, extremely wealthy and they are isolated from others. They are pursuing their own wealth and self-interest. Moreover, people

---

4 Ban Ki-Moon. Forward. *The Millennium Development Goals Report.* https://www.un.org/ millenniumgoals/2015_MDG_Report/pdf/MDG%202015%20rev%20(July%201).pdf.

5 Ibid., p. 8.

in poor countries have been getting poorer and people in rich countries have been getting richer.[6]

Recall that at COP-17, parties agreed to establish the Green Climate Fund that would assist poor countries to invest in renewable energy and to develop resilient strategies in dealing with climate change. At COP-18, which met in Doha in 2012, decisions were reached about compensation to poor countries that experienced grave harms from climatic events. The Sustainable Development Goals (SDGs) Summit met in New York City and on September 25, 2015, the Goals were adopted by all the world's nations to "cover nearly every aspect of our future—for our planet, and for humankind." Besides, it was said, "they concern all people, all countries, and all parts of society."[7]

Let us repeat, for emphasis. First of all the SDGs apply to all countries, not just to developing countries. Second, the SDGs reflect the understanding that the global community must tackle and reduce inequalities. Third, it was recognized that sustainable development is only possible if the pace of climate warming is slowed, and if all countries commit to ensuring that carbon dioxide emissions are reduced to zero by 2040–50. Only this would make it likely that global warming is reduced to 1.5 degrees Celsius.

Table 7.1 lists the SDGs.

Do note that before implementation each of the goals was discussed at the UN, in NGOs, and in local communities. These discussions helped to shape and elaborate each goal. We simply provide two examples in Box 7.1—specifically, reducing inequality (Goal 10) and combating climate change (Goal 13).

Taken together these ten targets, when implemented, would make an extraordinary difference, reducing the great inequalities between people living in rich countries and those living in poor countries, as well as the great inequalities within countries. Most of the targets are self-explanatory, but 10a, 10b, and 10c need clarification. Number 10a states that developing countries have yet to receive special and differential treatment in accordance with WTO agreements. This has been a problem that has dragged on and on and on,

---

6 Lant Pritchard. "Divergence Big Time": Policy Research Working Papers. The World Bank. http://elibrary.worldbank.org/doi/abs/10.1596/1813-9450-1522; Sudhir and Paul Segal, "The Distribution of Income," chapter 11 in A. B. Atkinson and F. Bourguignon, *Handbook of Income Distribution, Volume 2A* (Elsevier, Amsterdam, 2015): https://www.economics.ox.ac.uk/materials/papers/13376/anand-segal-handbook-pdf-mar15.pdf.

7 Learn about the SDGs: http://17goals.org./.

Table 7.1 Sustainable Development Goals, 2015–30

1) End poverty in all its forms everywhere

2) End hunger, achieve food security and improved nutrition, and promote sustainable agriculture

3) Ensure healthy lives and promote wellbeing for all at all ages

4) Ensure inclusive and equitable quality education and promote lifelong learning opportunities for all

5) Achieve gender equality and empower all women and girls

6) Ensure availability and sustainable management of water and sanitation for all

7) Ensure access to affordable, reliable, sustainable and modern energy for all

8) Promote sustained, inclusive and sustainable economic growth, full and productive employment, and decent work for all

9) Build resilient infrastructure, promote inclusive and sustainable industrialization, and foster innovation

10) Reduce inequality within and among countries

11) Make cities and human settlements inclusive, safe, resilient and sustainable

12) Ensure sustainable consumption and production patterns

13) Take urgent action to combat climate change and its impacts. Acknowledging that the United Nations Framework Convention on Climate Change is the primary international, intergovernmental forum for negotiating the global response to climate change.

14) Conserve and sustainably use the oceans, seas and marine resources for sustainable development

15) Protect, restore and promote sustainable use of terrestrial ecosystems, sustainably manage forests, combat desertification and halt and reverse land degradation, and halt biodiversity loss

16) Promote peaceful and inclusive societies for sustainable development, provide access to justice for all and build effective, accountable and inclusive institutions at all levels

17) Strengthen the means of implementation and revitalize the global partnership for sustainable development

Source: Sustainable Development Knowledge Platform: https://sustainabledevelopment. un.org/post2015/transformingourworld.

BOX 7.1 SDG GOAL 10. REDUCE INEQUALITY WITHIN AND AMONG COUNTRIES. TARGETS

10.1. By 2030, progressively achieve and sustain income growth of the bottom 40 per cent of the population at a rate higher than the national average

10.2. By 2030, empower and promote the social, economic and political inclusion of all, irrespective of age, sex, disability, race, ethnicity, origin, religion or economic or other status

10.3. Ensure equal opportunity and reduce inequalities of outcome, including by eliminating discriminatory laws, policies and practices and promoting appropriate legislation, policies and action in this regard

10.4. Adopt policies, especially fiscal, wage and social protection policies, and progressively achieve greater equality

10.5. Improve the regulation and monitoring of global financial markets and institutions and strengthen the implementation of such regulations

10.6. Ensure enhanced representation and voice for developing countries in decision-making in global international economic and financial institutions in order to deliver more effective, credible, accountable and legitimate institutions

10.7. Facilitate orderly, safe, regular and responsible migration and mobility of people, including through the implementation of planned and well-managed migration policies

10.a Implement the principle of special and differential treatment for developing countries, in particular least developed countries, in accordance with World Trade Organization agreements

10.b. Encourage official development assistance and financial flows, including foreign direct investment, to States where the need is greatest, in particular least developed countries, African countries, small island developing states and landlocked developing countries, in accordance with their national plans and programmes

10.c. By 2030, reduce to less than 3 per cent the transaction costs of migrant remittances and eliminate remittance corridors with costs higher than 5 per cent

Source: Sustainable Development Knowledge Platform: https://sustainabledevelopment. un.org/post2015/transformingourworld. The individual targets are renumbered here for clarity.

without being resolved.[8] Number 10b highlights the importance of official development assistance, or ODA, which is the percentage of gross national income that developed countries are expected to give to developing countries, and is currently set at 0.7 percent of the gross national income. Developed countries that meet this objective are: Sweden, United Arab Emirates, Norway, Luxembourg, Denmark, Netherlands, and United Kingdom.[9] And Number 10c refers to the transaction costs of remittances that migrants living and working in a host country must pay. Worldwide remittances in 2015 were $592 billion (in US dollars), and the transaction costs now paid by senders average 7.4 percent, more than double than that recommended in Goal 10c.[10]

Table 7.2 lists the targets for Goal 13 related to climate change. The Paris Climate Change Summit occurred just on the heels of the Sustainable Development Goals Summit (30 November–12 December and 25–27 September, respectively) and for that reason the targets are not detailed. However, they include a focus on assistance to developing countries by clarifying the importance of the Green Climate Fund (13.a) that assists developing countries acquire renewable energy technologies while 13.b reinforces 13.a and also highlights the special importance of small island states that are at peril.

THE PARIS AGREEMENT

It was clear when the SDGs were approved that the planet was heating at an unacceptable rate. In fact, scientists became gravely concerned about planetary heating in the 1970s (although it had been recognized already in the early

---

8 Aurelie Walker, "The WTO Has Failed Developing Nations": The *Guardian*. November 14, 2011: https://www.theguardian.com/global-development/poverty-matters/2011/nov/14/wto-fails-developing-countries. Aileen Kwa, "WTO and Developing Countries": Foreign Policy in Focus. http://fpif.org/wto_and_developing_countries/.

9 OECD. "Development Aid in 2015 Continues to Grow Despite Costs for in Donor Refugees. 2015 Preliminary ODA Figures": http://www.oecd.org/dac/stats/ODA-2015-detailed-summary.pdf.

10 The World Bank. "Remittances to Developing Countries Edge Up Slightly in 2015": April 13, 2016: http://www.worldbank.org/en/news/press-release/2016/04/13/remittances-to-developing-countries-edge-up-slightly-in-2015; Caroline Freund and Nikola Spatafora, "Remittances, Transaction Costs, and Informality": *Journal of Development Economics, 86* (2) June 2008: http://www.sciencedirect.com/science/article/pii/S0304387807000818.

Table 7.2 SDG Goal 13. Take urgent action to combat climate change and its impacts

13.1	Strengthen resilience and adaptive capacity to climate-related hazards and natural disasters in all countries
13.2	Integrate climate change measures into national policies, strategies and planning
13.3	Improve education, awareness-raising and human and institutional capacity on climate change mitigation, adaptation, impact reduction and early warning
13.a	Implement the commitment undertaken by developed-country parties to the United Nations Framework Convention on Climate Change to a goal of mobilizing jointly $100 billion annually by 2020 from all sources to address the needs of developing countries in the context of meaningful mitigation actions and transparency on implementation and fully operationalize the Green Climate Fund through its capitalization as soon as possible
13.b	Promote mechanisms for raising capacity for effective climate change-related planning and management in least developed countries and small island developing States, including focusing on women, youth and local and marginalized communities

Source: Sustainable Development Knowledge Platform: https://sustainabledevelopment.un.org/post2015/transformingourworld

nineteenth century). There was, however, a global response on December 12, 2015, when thousands gathered in Paris to adopt a treaty to hold the global average temperature to below two degrees Celsius above pre-industrial levels. Since then it has become clear that the objective must be 1.5 degrees. It seemed then and in the years of Obama's presidency that the world was united to stop, or slow, planetary warming, but a few months into his presidency, on June 1, 2017, Trump withdrew the United States from the treaty.[11] Besides emitting an extraordinary amount of CO_2, the United States does not collaborate with other rich countries in assisting poorer countries transition from fossil fuels to

---

11 Judith Blau, *The Paris Agreement: Climate Change, Solidarity, and Human Rights* (London: Palgrave Macmillan, 2017), https://www.palgrave.com/us/book/9783319535401; Judith Blau, *Crimes against Humanity: Climate Change and Trump's Legacy of Planetary Destruction* (New York: Routledge, 2019), https://www.routledge.com/Crimes-Against-Humanity-Climate-Change-and-Trumps-Legacy-of-Planetary/ Blau/p/book/9781138312685

renewable energy and a disproportionate number of Americans deny the Earth is warming. In any case, taking into account that some cities and states are trying to implement policies that promote renewable energy, America's response has been slow and largely underfunded.

CONCLUSION: THE WORLD'S TWO BIGGEST CHALLENGES

Global inequality has never been worse. According to one estimate, the richest 1 percent has more wealth than the rest of the world's population combined.[12] Another way of looking at this is to ask how many are poor. Almost half the world—over three billion people—live on less than $2.50 a day. At least 80 percent of the world's people live on less than $10 a day. Additionally, it is important to understand that for the most part the developed world caused the heating of the planet through industrialization, intensive agriculture, and imperial practices. This is why the Green Climate Fund is important. And, yet, more generally, one must be concerned that these horrific inequalities will greatly impair the degree of global cooperation that is required to slow the heating of the planet. Hungry people do not buy solar panels. Inequality and poverty are first-order global problems.

Yet facing all the world's peoples is the most horrific and comprehensive planetary disaster imaginable. Climate change will spare none, although it cannot be said that everyone will be equally affected. The Middle East will be intolerably hot. Many, if not most, of the small island states will be overcome by the sea. Few coastal cities will be spared. The poor who already suffer will suffer more. To be sure, climate change not only means warming. It also will mean the exacerbation of health problems and displaced populations, hunger, flooding, and violent storms. Because Trump gave the United States the license to abandon the constraints of the Paris Agreement and to abandon cooperation with the rest of the world, the United States will justifiably be held in contempt by the world's peoples.

It seems quite remarkable that every country in the world—except the United States—would team up to approve a treaty to limit the warming of the planet and end reliance on fossil fuels by 2050. That means countries that

---

12 Oxfam-Credit Suisse, "Global Wealth Report Again Exposes Inequality," November 22, 2016: https://www.oxfam.org/en/pressroom/reactions/credit-suisse-global-wealth-report-again-exposes-shocking-inequality.

otherwise do not respect many, if not most, international norms (such as North Korea) are party to the Paris Agreement and are taking steps to eliminate fossil fuels. It is not only in the self-interest of each country to cooperate, but in the collective interest as well. If the collective interest is not served, there will be collective disaster.

The spread of COVID-19 foregrounds and clarifies the deadly consequences of the climate crisis in ways that the United States never confronted before. The United States, which has withdrawn from the Paris Agreement, has also withdrawn from the World Health Organization at a time when global solidarity never meant so much.

CHAPTER 8

THE SDGS AND COVID-19

COVID-19 was first detected in December 2019 in Wuhan, China, and it rapidly became an international epidemic; it was declared a global pandemic by the UN's World Health Organization on 11 March 2020. The United Nations is playing a key international role in dealing with COVID-19—or shall we say, roles—since the UN's divisions engage the gamut of human experiences and concerns. Technically and ethically it became necessary for the United Nations to slow, modify, or stop some of its programs, when COVID-19 struck out, of great concern for human welfare. But happily, there were ceasefires declared in Syria and Yemen, saving many thousands of lives. Considerable effort and great concern have been the basis of the UN's response. In late March 2020 the UN issued a report for global distribution. The title is *Shared Responsibility, Global Solidarity: Responding to the Socio-Economic Impacts of Covid-19*. It begins:

> The United Nations family—and our global network of regional, sub-regional and country offices working for peace, human rights, sustainable development and humanitarian action—will support all governments, working with our partners, to ensure first and foremost that lives are saved, livelihoods are restored, and that the global economy and the people we serve emerge stronger from this crisis. More than ever before, we need solidarity, hope and the political will and cooperation to see this crisis through together.[1]

By June 2019 countries were well underway towards clarifying how they would achieve the sustainable development goals (SDGs).[2] However, six

---

1 United Nations. *Shared Responsibility, Global Solidarity*. March 2020: https://unsdg.un.org/sites/default/files/2020-03/SG-Report-Socio-Economic-Impact-of-Covid19.pdf.

2 *UN Officials Highlight Findings of 2019 SDG Progress Report:* https://sdg.iisd.org/news/un-officials-highlight-findings-of-2019-sdg-progress-report/.

months later it became clear that the SDGs needed to be scaled back because of the challenges posed by COVID-19. This chapter is based on the scaled-back (revised) SDGs, and instead of defining each SDG the challenges and possibilities are discussed.

REVISED SDGS, 2020[3]

GOAL 3: GOOD HEALTH AND WELL-BEING

The WHO is leading the global fight against COVID-19. WHO launched a campaign on March 23, 2020, which was, "Pass the message to kick out coronavirus," campaign by teaming up with FIFA, the international governing body of football.[4] At the press briefing, WHO Director-General Tedros Adhanom Ghebreyesus said that more than 300,000 cases of COVID-19 had been reported to WHO from almost every country in the world. It took 67 days from the first reported case to reach the first 100,000 cases, 11 days for the second 100,000 cases, and just 4 days for the third 100,000 cases. "You can't win a football game only by defending," he said. "To win, we need to attack the virus with aggressive and targeted tactics—testing every suspected case, isolating and caring for every confirmed case, and tracing and quarantining every close contact."[5]

GOAL 4: QUALITY EDUCATION

According to the United Nations Educational, Scientific and Cultural Organization (UNESCO), roughly 1.25 billion learners, or 72.9 percent of total enrolled learners worldwide, have been affected by the coronavirus outbreak as of 20 March. "In this crisis, which is first and foremost a public health crisis, our thoughts are of course with the sick and all those who are suffering today and

---

3 "UN Working to Fight Covid-19 and Achieve Global Goals": https://www.un.org/en/un-coronavirus-communications-team/un-working-fight-covid-19-and-achieve-global-goals.

4 World Health Organization. "Pass the Message to Kick Out Coronavirus": https://www.who.int/news-room/campaigns/connecting-the-world-to-combat-coronavirus/pass-the-message-to-kick-out-coronavirus.

5 World Health Organization. "Pass the Message. Five Steps to Kicking Out Coronavirus": https://www.who.int/news-room/detail/23-03-2020-pass-the-message-five-steps-to-kicking-out-coronavirus.

struggling against the coronavirus," says UNESCO Director-General Audrey Azoulay, adding that, "We must, however, remain mobilized, because this crisis also tells us several things that are very dear to UNESCO's mission." UNESCO is supporting governments for distance learning, scientific cooperation, and information support.

GOAL 5: GENDER EQUALITY

The United Nations Entity for Gender Equality and the Empowerment of Women (UN Women) has issued a checklist for COVID-19 response[6] that includes 10 requests for governments. Phumzile Mlambo-Ngcuka, UN under-secretary-general and UN women executive director, says that women carry countries' well-being on their shoulders and that right now, they are working day and night holding societies together, through health care, maternal care, elderly care, online teaching, child care, in pharmacies, in grocery stores, and as social workers. It is UN Women's job to support governments in upholding the rights of women and girls. "This is no less, and perhaps even more true, in times of crisis," she said. "The United Nations—and our global network of country offices—will support all Governments to ensure that the global economy and the people we serve emerge stronger from this crisis."[7]

GOAL 6: WATER AND SANITATION

One of the most effective ways to slow down transmission is to wash or sanitize hands. However, globally three billion people do not have access to even basic hand-washing facilities at home. Lack of access to clean water affects vulnerability to disease and ill-health. It is for this reason that UN-Water members and partners are committing to the Global Acceleration Framework,[8] which will unify the international community and deliver fast results in countries at an increased scale.

---

6 UN Women. "Checklist for Covid-19 Response": https://www.unwomen.org/en/news/stories/2020/3/news-checklist-for-covid-19-response-by-ded-regner.

7 UN Women. "Covid-19 Women Front and Centre": https://www.unwomen.se/covid-19-women-front-and-centre/.

8 UN, "Water to Develop Acceleration Framework for SDG-6": https://sdg.iisd.org/news/un-water-to-develop-acceleration-framework-for-SDG-6.

GOAL 8: DECENT WORK AND ECONOMIC GROWTH

On April 2, 2020, the ILO issued an international statement expressing alarm about unprotected workers:

> The novel coronavirus slowly spreads into the developing world, billions of people face a horrible choice: keep going to their informal jobs and risk contracting or spreading the virus, or stay at home and risk their family starving? Around 2 billion people—61% of the world's working population—toil in the informal economy—and have little or nothing to protect them if they're unable to go to work. They usually have neither sick pay nor health insurance and may not be eligible for government benefits given to furloughed workers. The situation threatens to be particularly acute in Africa, where 86% of employment is informal.[9]

GOAL 10: REDUCE INEQUALITIES

In crises, the most vulnerable, including women and children, people with disabilities, the marginalized, and the displaced, pay the highest price. At a press briefing recently, the secretary-general said that the Office of the United Nations High Commissioner for Refugees and the International Organization for Migration have been working hard to have a plan, working with the host countries to prevent the arrival of the virus in refugee camps or in settlements. He appealed for the international community to fully support those measures, which will be included in a $2 billion humanitarian appeal that the United Nations will launch soon. Filippo Grandi, United Nations high commissioner for refugees, expressed concern that measures adopted by some countries could block the right to seek asylum. "All states must manage their borders in the context of this unique crisis as they see fit. But these measures should not result in closure of avenues to asylum or of forcing people to return to situations of danger."[10]

---

9 International Labour Organization. "Coronavirus Imperils Billions of Informal Workers Especially Women"; https://qz.com/1831326/coronavirus-imperils-billions-of-informal-workers-especially-women/.

10 Statement by Filippo Grandi, High Commissioner for Refugees: https://www.unhcr.org/en-us/news/press/2020/3/5e7395f84/statement-filippo-grandi-un-high-commissioner-refugees-covid-19-crisis.html.

GOAL 16: PEACE, JUSTICE, AND STRONG INSTITUTIONS

This goal is about finding ways to make sure everyone lives in a peaceful society. The secretary-general today made an urgent appeal for an immediate global ceasefire in all corners of the world and for a united international effort to combat the pandemic ravaging the world. "The fury of the virus illustrates the folly of war," he said at a virtual press conference. "It is time to put armed conflict on lockdown and focus together on the true fight of our lives."[11]

GOAL 17: PARTNERSHIPS

To make all the goals a reality will require the participation of everyone, including governments, the private sector, civil society organizations, and people around the world. The fight against COVID-19 is no exception.

CONCLUSION

Until mid-March US president Donald Trump kept insisting that the United States had COVID-19—(the "flu") under control, and that it "would disappear shortly."[12] This explains why the United States lost two months in preparing for the onslaught, and it also helps to explain why there have been persistent shortages in the United States for testing kits, respirators, and other medical supplies. It also helps to explain why the gig economy—without worker security and good wages—has flourished in America and continued to flourish during the first months of 2020,[13] when other countries (except Brazil and maybe North Korea) were beginning to address COVID-19. Yes, Trump was criticized by scientists, but his comments were not intended to appeal to scientists, but to

---

11 "Covid-19. UN Chief Calls for Global Ceasefire to Focus on the True Fight for Our Lives." https://news.un.org/en/story/2020/03/1059972.

12 "Trump Wants US Opened up by Easter Despite Health Officials Warnings." https://www.nytimes.com/2020/03/24/us/politics/trump-coronavirus-easter.html.

13 Trump gave American companies the right to ignore pleas from their gig workers to have paid leaves from work. The gig economy is made up of three main components: the independent workers paid by the gig (i.e., a task or a project) as opposed to those workers who receive a salary or hourly wage; the consumers who need a specific service, for example, a ride to their next destination, or a particular item delivered; and the companies that connect the worker to the consumer in a direct manner: https://www.wired.com/story/covid-19-pandemic-aggravates-disputes-gig-work/.

the public—primarily and to reinforce the idea that American capitalism was vastly superior to that in other countries.

At the very beginning of the United Nations the United States played an active and constructive active role, such as assisting in peace and development efforts. Trump is a nationalist and denounces globalism. He pulled out of the Paris Agreement and threatened the UN with nonpayment of US dues. For a long time period in 2019 the United States refused to pay UN dues. Nor has it been helpful when the United States fails to provide dues and accompanies this with a meaningless formal reply. For example, the official American response to the UN when asked to provide resources to very poor countries to help advance the SDGs, the United States failed to reply.[14]

Yes, it is the case that the United State is one of the richest countries in the world and boasts the oldest democracy as well as extraordinary ethnic, cultural, and racial diversity. It has played a largely constructive role in the United Nations, and though there have been times when this is not the case, never ever has the United States been so consistently obstructionist as now.

---

14 "While China Embraces the SDGs, the US Government Would Rather Not Talk about Them": https://www.devex.com/news/while-china-embraces-the-sdgs-the-us-government-would-rather-not-talk-about-them-95597.

CHAPTER 9

TURNING THE PAGE

The entire world is now threatened by two horrific processes—(1) a growing epidemic—COVID-19—with fatal consequences and (2) an unacceptable level of heating. The first is not well understood but is extremely frightening— it is capricious and dangerous. But one can say that it is so scary and frightening that with support from the United Nations over a dozen countries are entering into peace talks, so they can devote their time and resources to ending COVID-19.[1] Yet (2)—(planetary heating) is the more dangerous, since the entire world and its people are at risk with a possible five degree Centigrade increase in heat by the end of the century. We will discuss (2) in this chapter and (1) in the next and final chapter.

The agreed upon deadline to slow the heating of the planet is 2050, when we collectively together end our reliance on fossil fuels and—one hopes— embrace solar, wind, and waves. The reason is that fossil fuels—that is, coal, oil, and natural gas—must be replaced by renewable energy because the burning of fossil fuels creates carbon dioxide, which being a greenhouse gas, drives a heating effect that warms the planet. It is not an easy revolution, but it is guided by a scientific consensus that the burning of fossil fuels will destroy the Earth and that renewable energy—solar, wind, tides— has certain benefits, such as sustaining good air quality.[2] And the revolution is enthusiastically backed by young people who are adamant, determined, and far more aware of impending climate change than adults.

---

1 United Nations. Update on the Secretary General's Appeal for a Ceasefire, April 2, 2020: https://www.un.org/sites/un2.un.org/files/update_on_sg_appeal_for_ceasefire_april_2020.pdf; /.
2 Union of Concerned Scientists. "Benefits of Renewable Energy Use": December 20, 2017: https://www.ucsusa.org/resources/benefits-renewable-energy-use.

In this chapter we will briefly compare fossil fuel with renewable energy, highlight what rich countries are doing to work with developing countries to adopt renewable energy technology, and the proposed legislation in the United States to advance renewable energy and accompanying programs.

FOSSIL FUEL VERSUS RENEWABLE ENERGY

Fossil fuel is extracted from buried combustible geologic deposits of organic materials, formed from decayed plants and animals that have been converted to petroleum, coal, or natural gas by exposure to heat and pressure in the Earth's crust over hundreds of millions of years. The burning of fossil fuels by humans is the largest source of emissions of carbon dioxide, which is one of the greenhouse gases that contribute to global warming. In contrast, renewable energy resources capture their energy from existing flows of energy, from ongoing natural processes, such as sunshine, wind, flowing water (notably tides), biological processes, and geothermal heat flows.[3] In other words, renewable energy is from an energy resource that is replaced rapidly by a natural process such as power generated from the Sun or from the wind.

Most renewable forms of energy, other than geothermal and tidal power, ultimately come from the Sun, and they are solar, wind, water, geothermal, bioenergy (and, taking into account problems of safety, nuclear). Renewable energy is increasing dramatically all around the globe,[4] and the deadline set by UN member states to adopt clean—renewable—energy across the board is 2030.[5] The United States poses a particular problem—indeed a global one—because Trump withdrew the United States from the Paris Agreement, which is the international treaty for reducing and ending reliance on fossil fuels.[6] Because

---

3 "Fossil fuel." *Science Daily:* https://www.sciencedaily.com/terms/fossil_fuel.htm

4 UN Environment Programme: "A Decade of Renewable Energy Investment, Led by Solar Tops US Dollar 25 Trillion," September 5, 2019: https://www.unenvironment.org/news-and-stories/press-release/decade-renewable-energy-investment-led-solar-tops-usd-25-trillion; United Nations University. "UN Launches Decade-long Sustainable Energy for All Initiative." 2014•04•10: https://ourworld.unu.edu/en/un-launches-decade-long-sustainable-energy-for-all-initiative.

5 UN. Sustainable Development Goals. Goal 7 Affordable and Clean Energy. https://sdg-tracker.org/energy.

6 "Statement by President Trump on the Paris Climate Accord, June 1, 2017. https://www.whitehouse.gov/briefings-statements/statement-president-trump-paris-climate-accord/.

the US economy is huge and American economic actors are powerful and global, Trump's decision affects the entire planet.[7]

INTERNATIONAL GREEN CLIMATE FUND

The Green Climate Fund (GCF) is a new global fund created to support the efforts of developing countries to respond to the challenge of climate change. GCF helps developing countries limit or reduce their greenhouse gas emissions and adapt to the warming planet. It seeks to promote a paradigm shift to low-emission and climate-resilient development taking into account the needs of nations that are particularly vulnerable to climate change impacts.

It was set up by the 194 countries who are parties to the United Nations Framework Convention on Climate Change (UNFCCC) in 2010, as part of the Convention's financial program. It aims to deliver equal amounts of funding to mitigation and adaptation, while being guided by the Convention's principles and provisions. Forty-nine countries have contributed to the GCF, and several countries receive funds annually. It is fair and appropriate that rich, developed countries help developing countries, given their long histories of exploitation.

GREEN DEAL (EUROPE)

The European plan is comprehensive and includes 10 main points: (1) The EU will reach net-zero greenhouse gas emissions by 2050; (2) Circular economy (discussed in Chapter 4); (3) Building renovation; (4) Promote pollution-free environment; (5) Increase biodiversity; (6) Green and healthier agriculture; (7) Strengthen emission standards and encourage electric cars; (8) Assist member states that need assistance; (9) Encourage innovation; (10) Work with other non-EU countries to increase climate ambitions.

THE GREEN NEW DEAL (UNITED STATES)

In February 2019, Senator Edward Markey and Representative Alexandria Ocasio-Cortez proposed a resolution for an American Green New Deal. It calls for a 10-year national mobilization whose primary goals would be:[8]

---

7 Judith Blau, *Climate Change and Trump's Legacy of Planetary Destruction.* (London: Routledge, 2018.)
8 https://www.nti.org/about/projects/NTI-WHO-Emergency-Response/ /https://www.nti.org/about/projects/NTI-WHO-Emergency-Response/.

Guaranteeing a job with a family-sustaining wage, adequate and medical leave, paid vacations, and retirement security to all people of the United States.

Providing all people of the United States with – (i) high-quality health care; (ii) affordable, safe, and adequate housing; (iii) economic security; and (iv) access to clean water, clean air, healthy and affordable food, and nature.

Providing resources, training, and high-quality education to all people of the United States.

Meeting 100 percent of the power demand in the United States through clean, renewable, and zero-emission energy sources.

Repairing and upgrading the infrastructure in the United States, including by eliminating pollution and greenhouse gas emissions as much as possible.

Building or upgrading to energy-efficient power grids, and working to ensure affordable access to electricity.

Upgrading all existing buildings in the United States and building new buildings to achieve maximal energy efficiency, water efficiency, safety, affordability, comfort, and durability, including through electrification.

Overhauling transportation systems in the United States to eliminate pollution and greenhouse gas emissions from the transportation sector as much as feasible, including through investment in—(i) zero-emission vehicle infrastructure and manufacturing; (ii) clean, affordable, and accessible public transportation; and (iii) high-speed rail.

Spurring massive growth in clean manufacturing in the United States and removing pollution and greenhouse gas emissions from manufacturing and industry as much as is technologically feasible.

Working collaboratively with farmers and ranchers in the United States to eliminate pollution and greenhouse gas emissions from the agricultural sector as much as is technologically feasible.

There is no question that if adopted it would slow the rise of heat, reduce economic inequalities, ensure that there were infrastructure upgrades, create jobs, tackle pollution, repair roads and bridges, and enhance social and economic security for Americans.

GREEN NEW DEAL FOR PUBLIC HOUSING ACT

A few months later, in November 2019, Representative Alexandria Ocasio-Cortez of New York and Senator Bernie Sanders of Vermont announced they would introduce the Green New Deal for Public Housing Act that would take on both the housing crisis and the climate crisis, and create jobs. The proposed legislation would cut 5.6 million tons of carbon emissions, create 240,000 jobs, and overhaul one million public housing units. The idea is to tackle several crises at once.[9]

CONCLUSION

It is important to recognize that because President Trump pulled out of the Paris Agreement the federal government is under no pressure to work collaboratively with other countries to slow or stop the heating of the planet; nor does it have any incentive to support renewable energy. Whatever state governments, companies, towns, and Americans do to advance and support renewable energy is entirely based on their own initiative and has no formal backing by the federal government. Thousands of scientists appealed to the world's peoples in a letter in November 2019 to warn them of the overheating planet. The letter stated, "We declare clearly and unequivocally that planet Earth is facing a climate emergency." "To secure a sustainable future, we must change how we live. [This] entails major transformations in the ways our global society functions and interacts with natural ecosystems."[10]

But to live in a country that officially does not formally or officially recognize the dangers of accumulating carbon dioxide emissions, it is easy to understand why there are so many Americans who are indifferent or are not informed. Greta Thunberg would say, "Listen to the scientists."

---

9 Senator Bernard Sanders. Green New Deal for Public Housing Act. 11/14/2019. S.2876—116th Congress (2019–20): https://www.congress.gov/bill/116th-congress/senate-bill/2876.

10 William J. Ripple, Christopher Wolf, Thomas M. Newsome, William J. Ripple, Christopher Wolf, Thomas M. Newsome, Phoebe Barnard, William R. Moomaw, "World Scientists Warning of a Climate Emergency." *BioScience*. Vol. 70, 1: January 2020, Page 100. https://doi.org/10.1093/biosci/biz152; list of 11,258 scientist signatories from 153 countries linked to article. https://academic.oup.com/bioscience/advance-article/doi/10.1093/biosci/biz088/5610806.

CHAPTER 10

CONCLUSIONS: TO END CAPITALISM

You have stolen my dreams and my childhood with your empty words. And yet I'm one of the lucky ones. People are suffering. People are dying. Entire ecosystems are collapsing. We are in the beginning of a mass extinction, and all you can talk about is money and fairy tales of eternal economic growth. How dare you! – Greta Thunberg, Sept 23, 2019[1]

Dear friends of planet Earth, … I have asked you here to sound the alarm. Climate change is the defining issue of our time—and we are at a defining moment. We face a direct existential threat. Climate change is moving faster than we are—and its speed has provoked a sonic boom SOS across our world.

António Guterres,[2] Sept 10, 2018

The dire prediction is that the Earth will not be habitable as heating accelerates. The prediction made by scientists is that Earth will heat somewhere between three and four degrees Centigrade by 2100. Let us be clear: (1) The source of global heating was—and continues to be—capitalism, fueled by greed and exploitation.[3] (2) Capitalism originated in America with slavery. It was fine-tuned in America and then outsourced to colonies and then to the entire Third

---

1 Transcript: Greta Thunberg's Speech at the U.N. Climate Action Summit, September 23, 2019: https://www.npr.org/2019/09/23/763452863/transcript-greta-thunbergs-speech-at-the-u-n-climate-action-summit.

2 António Guterres, "A Sustainable Global Economy Must Arise Once COVID-19 Pandemic Is Reversed," UN chief tells G-20 summit: https://news.un.org/en/story/2020/03/1060442.

3 Naomi Klein, "Capitalism vs. the Climate," *The Nation,* November 9, 2011: https://www.thenation.com/article/archive/capitalism-vs-climate/.

World and developing countries. What are the instances? (3) These are: every mine, oil well, colonial expansion, act of imperialism or expansion of trade, and act of exploitation of land and Earth's resources.

According to journalist Gia Vance,

> To slow heating must be the objective. Food production will need to be more intensive, efficient and industrial. This will be a mostly vegetarian world, largely devoid of fish and without the grazing area or resources for livestock. Poultry may be viable on the edges of farmland and synthetic meats and other foods will meet some of the demand. Heat-tolerant, drought-resistant crop varieties, such as cassava and millet, will replace many of our current unmodified staples such as rice and wheat and they will grow faster and with greater water efficiency because of the high CO_2 levels.[4]
>
> Our best hope lies in cooperating as never before to radically reorganize our world: decoupling the political map from geography. However unrealistic it sounds, we'd need to look at the world afresh and see it in terms of where the resources are and then plan the population, food and energy production around that. It would mean abandoning huge tracts of the globe and moving Earth's human population to the high latitudes: Canada, Siberia, Scandinavia, parts of Greenland, Patagonia, Tasmania, New Zealand, and perhaps newly ice-free parts of the western Antarctic coast.

Some propose socialism, others propose worker cooperatives, like Mondragon Corporation, and still others propose small independent communities. None of the 99 percent who find themselves poor, or relatively poor, would defend capitalism. We focus on the World Health Organization (WHO) that is on the front line of the anti-COVID-19 battle—carrying out research, coordinating scientific projects, working with all countries. Their very core or heart or belief is to protect the world's peoples.

WORLD HEALTH ORGANIZATION

In January 2020 the WHO declared COVID-19 a global epidemic and on March 11, 2020, a pandemic. It set up several funds:1) the Solidarity

---

4 Gia Vance, The *Guardian,* "The Heat Is on over Climate Change: Only Radical Measures Will Work": https://www.theguardian.com/environment/2019/may/18/climate-crisis-heat-is-on-global-heating-four-degrees-2100-change-way-we-live.

Fund—which allows the WHO to dispatch teams of scientists and practitioners to the site of disease outbreaks around the globe within 24 hours;[5] 2) the Healthy at Home Fund; and 3) a research fund—"Blueprint" or rapid assessment of local research trials and practice to see if they are generalizable.[6] In Box 10.1 are excerpted phrases from the speeches of Tedros Adhanom Ghebreyesus ("Tedros"), director general of the WHO. The speeches cover technical, medical, and scientific concerns.

Apparently, in response to Tedros daily speeches and his call for global solidarity, Trump announced that the United States would leave WHO and withdraw US funding.[7] The G-7 refused to back Trump and the UN defended Tedros. WHO is continuing with its scientific work. Besides, American comedians have booked time on TV to defend the WHO against Trump's withdrawal of funds.[8]

But let's step back to understand what is going on. Yes, it is relevant that Trump's financial self-interest is bound up with the policies he promotes. For example, he has financial interest in hydroxychloroquine—the medicine he is promoting as a cure for COVID-19, and which WHO can find no support for, or that is successful in curing COVID-19. A recent issue of *Newsweek* documents 2,500 instances in which Trump has entrenched corruption.[9] He owns properties in 25 countries, which influences his decision making.[10]

5　Global Emergency Response Fund: https://www.nti.org/about/projects/NTI-WHO-Emergency-Response/.

6　Blueprint. https://www.who.int/blueprint/about/en/.

7　"Donald Trump Threatens to Withdraw Funding from 'China-centric' WHO" SBS News:https://www.sbs.com.au/news/donald-trump-threatens-to-withdraw-funding-from-china-centric-who.

8　Glenn Kessler, Trump's False Claim: "The president's announcement that he would suspend payments to the World Health Organization in the midst of the coronavirus pandemic contained a number of false or misleading claims. He faulted the WHO for believing that China was doing a good job and praising its transparency—when he had done the same thing at the time. He claimed the WHO 'fought' his decision to impose some restrictions on travel from China, but WHO officials said nothing publicly; opposition to travel restrictions has been a consistent WHO policy": https://www.washingtonpost.com/politics/2020/04/17/trumps-false-claim-that-who-said-coronavirus-was-not-communicable/.

9　Jessica Kwong. "Trump Has More Than 2,500 Conflicts of Interest and Counting, Live Tracker by Watchdog Finds" Newsweek.com: https://www.newsweek.com/trump-conflicts-interest-tracker-1466800.

10　Anna Massoglia and Karl Evers-Hillstrom, "Worlds of Influence": https://www.opensecrets.org/news/2019/06/trump-foreign-business-interests/.

BOX 10.1 THE DIRECTOR-GENERAL OF THE WORLD HEALTH ORGANIZATION—TEDROS ADHANOM GHEBREYESUS—HIGHLIGHTS SOLIDARITY IN HIS SPEECHES IN MARCH 2020

We are interdependent. We cannot win without **Solidarity.**[a]

This problem can only be solved with international cooperation and international **Solidarity.**[b]

That means a paradigm shift in global **Solidarity**—in sharing experiences, expertise and resources, and in working together to keep supply lines open, and supporting nations who need our support.[c]

I want to begin by reiterating the Secretary-General's comments that now is the time for **Solidarity** in the face of this threat to all of humanity.[d]

We are so grateful that several countries have sent Emergency Medical Teams to care for patients and train health workers in other countries that need support. This is an incredible example of international **Solidarity**. But it's not an accident[e].

As I keep saying, Solidarity is the key to defeating COVID-19—**Solidarity** between countries, but also between age groups. Thank you for heeding our call for **Solidarity, Solidarity, Solidarity**.[f]

We have confidence in our Member States, and the only way we can defeat this pandemic, as we have always been saying, is through Solidarity. Solidarity, Solidarity, Solidarity.[g]

That's what we're seeing now. This is a common enemy. Let's keep that Solidarity up. We're one human race, and that suffices actually. This is an invisible enemy against humanity.[h]

This amazing spirit of human Solidarity must become even more infectious than the virus itself. Although we may have to be physically apart from each other for a while, we can come together in ways we never have before.[i]

We have encouraged them to unite, stressing that no country can fight this alone, and calling on all countries to build on the Solidarity already sparked by the crisis.[j]

We have called this study the SOLIDARITY trial. The SOLIDARITY trial provides simplified procedures to enable even hospitals that have been overloaded to participate.[k]

That's why we keep talking about Solidarity. This is not just a threat for individual people, or individual countries. We're all in this together, and we can only save lives together.[1]

---

[a] WHO Director-General's Remarks for G20 Trade Ministers: March 30, 2020: https://www.who.int/dg/speeches/detail/who-director-generals-remarks-for-g20-trade-ministers.

[b] WHO Director-General's Opening Remarks at the Media Briefing on COVID-19—March 27, 2020: https://www.who.int/dg/speeches/detail/who-director-general-s-opening-remarks-at-the-media-briefing-on-covid-19-27-march-2020.

[c] WHO Director-General's Remarks at the G20 Extraordinary Leaders' Summit on COVID-19—March 26, 2020: https://www.who.int/dg/speeches/detail/who-director-general-s-remarks-at-the-g20-extraordinary-leaders-summit-on-covid-19-26-march-2020.

[d] WHO Director-General's Remarks Launch of Appeal: Global Humanitarian Response Plan—March 25, 2020: https://www.who.int/dg/speeches/detail/who-director-general-s-remarks-launch-of-appeal-global-humanitarian-response-plan-25-march-2020.

[e] WHO Director-General's opening remarks at the media briefing on COVID-19—March 23, 2020. https://www.who.int/dg/speeches/detail/who-director-general-s-opening-remarks-at-the-media-briefing-on-covid-19-23-march-2020.

[f] ibid.

[g] Reuters. March 20, 2020 "WHO Message to Youth on Coronavirus: 'You are not invincible'": https://www.reuters.com/article/us-health-coronavirus-who/who-message-to-youth-on-coronavirus-you-are-not-invincible-idUSKBN21733O.

[h] WHO Director-General's opening remarks at the Mission briefing on COVID-19—March 19, 2020: https://www.who.int/dg/speeches/detail/who-director-general-s-opening-remarks-at-the-mission-briefing-on-covid-19-19-march-2020.

[i] WHO Director-General's opening remarks at the media briefing on COVID-19—April 15, 2020: https://www.who.int/dg/speeches/detail/who-director-general-s-opening-remarks-at-the-media-briefing-on-covid-19-15-april-2020.

[j] WHO Director-General calls on G20 to Fight, Unite, and Ignite against COVID-19 https://www.who.int/news-room/detail/26-03-2020-who-s-director-general-calls-on-g20-to-fight-unite-and-ignite-against-covid-19.

[k] WHO Director-General's opening remarks at the media briefing on COVID-19—March 18, 2020: https://www.who.int/dg/speeches/detail/who-director-general-s-opening-remarks-at-the-media-briefing-on-covid-19-18-march-2020.

[l] WHO Director-General's opening remarks at the media briefing on COVID-19—March 5, 2020: https://www.who.int/dg/speeches/detail/who-director-general-s-opening-remarks-at-the-media-briefing-on-covid-19-5-march-2020.

SOLIDARITY, NOT CAPITALISM

But the point we want to make does not have to do with Trump as a corrupt American president, but with capitalism itself. We want you to remember what Tedros tells the world everyday—Solidarity!! We have also followed Oxfam's reports that they present annually at the World Economic Forum. First, here is the basis of what Oxfam presented in 2019:

Billionaire fortunes increased by 12 percent last year—or $2.5 billion a day —while the 3.8 billion people who make up the poorest half of humanity saw their wealth decline by 11 percent.[11]

Here is a summary of what Oxfam presented in 2020.[12]

1. The richest 1% in the world have more than double the wealth of 6.9 billion people
2. The world's richest 22 men have more money than all the women in Africa
3. Women and girls put in 12.5 billion hours of unpaid work every day
4. Women's unpaid care work has a monetary value of $10.8 trillion a year

Oxfam notes,

Extreme inequality is out of control. Hundreds of millions of people are living in extreme poverty while huge rewards go to those at the very top. There are more billionaires than ever before, and their fortunes have grown to record levels. Meanwhile, the world's poorest got even poorer. Many governments are fueling this inequality crisis. They are massively under taxing corporations and wealthy individuals, yet underfunding vital public services like healthcare and education.

These policies hit the poor hardest. The human costs are devastating, with women and girls suffering the most. Despite their huge contribution to our

---

11 "Billionaire Fortunes Grew by $2.5 Billion a Day Last Year as Poorest Saw Their Wealth Fall": https://www.oxfam.org/en/press-releases/billionaire-fortunes-grew-25-billion-day-last-year-poorest-saw-their-wealth-fall.
12 World Economic Forum: https://www.weforum.org/agenda/2020/01/5-shocking-facts-about-inequality-according-to-oxfam-s-latest-report/.

societies through unpaid care work, they are among those who benefit the least from today's economic system.[13]

And we might add that capitalism has become a cruel, cruel joke, and as Oxfam puts it, "perpetrating myths about talent, skill, qualifications and mental ability[...]. Those at the top of the ladder are not smarter, more agile, or stronger than those on the bottom. They are simply luckier, more competitive, with more far-flung social and economic networks."[14] We would put it stronger than Oxfam: capitalism is corrupt, corrupting, dishonest, fraudulent. It creates horrific inequalities that are without merit, empirical basis, or grounding whatsoever.

Yes, one way to end capitalism is to join together in solidarity, as the Head of the UN WHO advocates. In fact the United Nations —of which WHO is a part—has always rested on the joint principles of Equality and Solidarity.[15] On the first page of the Charter is this sentence: "WE THE PEOPLES OF THE UNITED NATIONS DETERMINE to save succeeding generations from the scourge of war, which twice in our life-time has brought untold sorrow to mankind, and to reaffirm faith in fundamental human rights, in the dignity and worth of the human person, in the equal rights of men and women and of nations large and small."[16] Article 1 says this: "To develop friendly relations among nations based on respect for the principle of equal rights and self-determination of peoples, and to take other appropriate measures to strengthen universal peace." Article 2 states this: "The Organization is based on the principle of the sovereign equality of all its Members."

So yes, we propose that the United Nations be a model for a New World Order without capitalism and without inequality. A very, very brief history will suggest why we conclude this:

---

13 https://www.oxfam.org/en/5-shocking-facts-about-extreme-global-inequality-and-how-even-it.

14 https://www.pnas.org/content/113/47/13354.full; https://money.com/are-millionaires-smarter-than-the-rest-of-us/; https://www.thejournal.ie/readme/opinion-why-do-rich-people-lie-cheat-and-steal-more-than-those-on-low-incomes-4647197-May2019/; https://www.aljazeera.com/indepth/opinion/coronavirus-signal-capitalism-200330092216678.html.

15 https://www.un.org/en/sections/about-un/un-card-10-facts/index.html; https://treaties.un.org/doc/publication/ctc/uncharter.pdf.

16 https://treaties.un.org/doc/publication/ctc/uncharter.pdf.

In 1945 about 60 million people were dead; there were millions of displaced and homeless people, when in Germany, it has been estimated, 70% of housing had gone and, in the Soviet Union, 1,700 towns and 70,000 villages. Factories and workshops were in ruins, fields, forests and vineyards ripped to pieces. Millions of acres in north China were flooded after the Japanese destroyed the dykes. Many Europeans were surviving on less than 1,000 calories per day; in the Netherlands they were eating tulip bulbs. [...] Britain had largely bankrupted itself fighting the war and France had been stripped bare by the Germans. They were struggling to look after their own peoples and deal with reincorporating their military into civilian society. The four horsemen of the apocalypse—pestilence, war, famine and death—so familiar during the middle ages, appeared again in the modern world.[17]

In other words, homelessness, hunger, and misery dominated the world in 1945, when the United Nations was created to end all of these. It has certainly tried to end all three and has been somewhat successful but nation-states in alliance with capitalism have nurtured—yes, nurtured—homelessness, hunger, and misery. Amazingly the United Nations does not embrace capitalism. Instead it embraces equality (using money simply as a tool—to pay staff and underwrite programs), which is to say the UN promotes and underwrites gender equality, racial equality, equal rights to an identity, equal rights to education, and equal rights to opportunity. Yes, the United Nations was the birthplace of human rights. In these regards, the UN has been a model and always will.

Nation-states, however, are being influenced by new technologies: the circular economy and blockchain technology. They are providing us with early models of equal rights and equal voice. States, of course, will be different. Some will be capitalist, and some will be socialist. Some will be partisan and some not. Some will have a religious identity and others not. Evidence suggests that authoritarianism is slowly giving way to democracy. And there are plenty of examples of grounded participatory democracy, including participatory budgeting in New York City, participatory budgeting in Porto Alegre, participatory democracy in Spain, participatory budgeting in Cambridge, Massachusetts,

---

17 https://www.theguardian.com/world/2009/sep/11/second-world-war-rebuilding.

comprehensive participatory democracy in the Basque country, Spain. There is remarkable happiness that accompanies participatory democracy.[18]

Yes, all states and all peoples deserve equal rights and equal voice

Yes, that would be Solidarity.

There would be a global ceasefire, global development, global recognition, global equality.

---

18 Judith M Green. *Deep Democracy*. (Lanham, Maryland: Rowman & Littlefield, 1999); Dimitrios Roussopoulos, C. George Benello (editors), *Participatory Democracy* (Montreal: Black Rose Books, 2005).

GLOSSARY: TERMS RELEVANT FOR (1) GLOBAL WARMING, AND (2) COVID-19

(1) GLOBAL WARMING

Aerosols (Atmospheric) Are fine solid or liquid particles, caused by people or occurring naturally, that are suspended in the atmosphere. Aerosols can cause cooling by scattering incoming radiation or by affecting cloud cover. Aerosols can also cause warming by absorbing radiation.

Algal bloom A sudden, rapid growth of algae in lakes and coastal oceans caused by a variety of factors including, for example, warmer surface waters or increased nutrient levels. Some algal blooms may be toxic or harmful to humans and ecosystems.

Anthropocene Epoch Is an unofficial unit of geologic time, used to describe the most recent period in Earth's history when human activity started to have a significant impact on the planet's climate and ecosystems.

Atmospheric aerosols Microscopic particles suspended in the lower atmosphere that reflect sunlight back to space. These generally have a cooling effect on the planet and can mask global warming. They play a key role in the formation of clouds, fog, precipitation, and ozone depletion in the atmosphere.

Biofuel Fuel produced from plant or animal matter such as corn or manure.

Cap and trade An emission trading scheme whereby businesses or countries can buy or sell allowances to emit greenhouse gases via an exchange. The volume of allowances issued adds up to the limit, or cap, imposed by the authorities.

Carbon capture and storage The collection and transport of concentrated carbon dioxide gas from large emission sources, such as power plants. The gases are then injected into deep underground reservoirs. Carbon capture is sometimes referred to as geological sequestration.

Carbon dioxide (CO_2) Carbon dioxide is a gas in the Earth's atmosphere. It occurs naturally and is also a byproduct of human activities, such as burning fossil fuels. It is the principal greenhouse gas produced by human activity.

Carbon sequestration The process of storing carbon dioxide. This can happen naturally, as growing trees and plants turn CO_2 into biomass (wood, leaves, and so on). It can also refer to the capture and storage of CO_2 produced by industry. See Carbon capture and storage.

Carbon sink Any process, activity, or mechanism that removes carbon from the atmosphere. The biggest carbon sinks are the world's oceans and forests, which absorb large amounts of carbon dioxide from the Earth's atmosphere.

CO_2 A heavy colorless gas that does not support combustion, dissolves in water to form carbonic acid, and is formed especially in animal respiration and in the decay or combustion of animal and vegetable matter, is absorbed from the air by plants in photosynthesis, and is used in the carbonation of beverages. A key greenhouse gas that drives global climate change, it continues to rise every month.

COP Conference of the parties (COP; French: Conférence des Parties, CP) is the governing body of an international convention. Conventions with a COP include: United Nations Framework Convention on Climate Change.

Deforestation The permanent removal of standing forests that can lead to significant levels of carbon dioxide emissions.

Ecocide Is a crime against the living natural world—ecosystem loss, damage, or destruction is occurring every day.

Ecosystem All the living things in a particular area as well as components of the physical environment with which they interact, such as air, soil, water, and sunlight.

El Niño-Southern Oscillation A natural variability in ocean water surface pressure that causes periodic changes in ocean surface temperatures in the tropical Pacific ocean. El Niño Southern Oscillation (ENSO) has two phases: the warm oceanic phase, El Niño, accompanies high air surface pressure in the western Pacific, while the cold phase, La Niña, accompanies low air surface pressure in the western Pacific. Each phase generally lasts for 6 to 18 months. ENSO events occur irregularly, roughly every 3 to 7 years. The extremes of this climate pattern's oscillations cause extreme weather (such as floods and droughts) in many regions of the world.

Feedback loops In climate change, a feedback loop is the equivalent of a vicious or virtuous circle—something that accelerates or decelerates a warming trend. A positive feedback accelerates a temperature rise, whereas a negative feedback decelerates it.

Fossil fuels Natural resources, such as coal, oil, and natural gas containing hydrocarbons. These fuels are formed in the Earth over millions of years and produce carbon dioxide when burnt.

G77 The main negotiating bloc for developing countries, allied with China (G77+China). The G77 comprises 130 countries, including India and Brazil, most African countries, the grouping of small island states (Aosis), the Gulf states, and many others, from Afghanistan to Zimbabwe.

Greenhouse effect The greenhouse effect causes the atmosphere to retain heat. Greenhouse gases (GHGs) like water vapor (H_2O), carbon dioxide (CO_2), and methane (CH_4) absorb energy, slowing or preventing the loss of heat to space. In this way, GHGs act like a blanket, making Earth warmer than it would be.

IPCC The Intergovernmental Panel on Climate Change is a scientific body established by the United Nations Environment Programme and the World Meteorological Organization. It reviews and assesses the most recent scientific, technical, and socioeconomic work relevant to climate change, but does not carry out its own research. The IPCC was honored with the 2007 Nobel Peace Prize.

IRENA The International Renewable Energy Agency is an intergovernmental organization mandated to facilitate cooperation, advance knowledge, and promote the adoption and sustainable use of renewable energy.

La Niña The direct opposite of El Niño, occurs when sea surface temperatures in the central Pacific Ocean drop to lower-than-normal levels. The cooling of this area of water near the equator, which typically unfolds during late fall into early winter, yields impacts around the globe.

LDCs Least Developed Countries represent the poorest and weakest countries in the world. The current list of LDCs includes 49 countries—33 in Africa, 15 in Asia and the Pacific, and 1 in Latin America.

Methane This is the second most important man-made greenhouse gas. Sources include both the natural world (wetlands, termites, wildfires) and human activity (agriculture, waste dumps, leaks from coal mining).

Mitigation Measures to reduce the amount and speed of future climate change by reducing emissions of heat-trapping gases or removing carbon dioxide from the atmosphere.

Ocean acidification The process by which ocean waters have become more acidic due to the absorption of human-produced carbon dioxide, which interacts with ocean water to form carbonic acid and lower the ocean's pH. Acidity reduces the capacity of key plankton species and shelled animals to form and maintain shells.

Ozone A colorless gas consisting of three atoms of oxygen, readily reacting with many other substances. Ozone in the upper atmosphere protects the Earth from harmful levels of ultraviolet radiation from the Sun. In the lower atmosphere ozone is an air pollutant with harmful effects on human health.

Paris Agreement At COP 21 in Paris, on December 12, 2015, parties to the UNFCCC reached a landmark agreement to combat climate change and to accelerate and intensify the actions and investments needed for a sustainable low carbon future. The Paris Agreement—for the first time—brings all nations into a common cause to undertake ambitious efforts to combat climate change and adapt to its effects, with enhanced support to assist developing countries to do so. As such, it charts a new course in the global climate effort.

Phytoplankton Microscopic organisms (single-celled plants, bacteria, and protists) capable of photosynthesis. Phytoplankton are found in oceans, seas, and freshwater, and are an essential component of aquatic ecosystem.

Renewable energy It is energy created from sources that can be replenished in a short period of time. The five renewable sources used most often are: biomass (such as wood and biogas), the movement of water, geothermal (heat from within the Earth), wind, and solar.

Risk assessment Studies that estimate the likelihood of specific sets of events occurring and their potential positive or negative consequences.

SDGs The Sustainable Development Goals, also known as the Global Goals, were adopted by all United Nations Member States in 2015 as a universal call of action to end poverty, protect the planet, and ensure that all people enjoy peace and prosperity by 2030.

Sink A natural or technological process that removes carbon from the atmosphere and stores it.

Tipping point A tipping point is a threshold for change, which, when reached, results in a process that is difficult to reverse. Scientists say it is urgent that policy makers halve global carbon dioxide emissions over the next 50 years or risk triggering changes that could be irreversible.

UNFCCC The United Nations Framework Convention on Climate Change is an international environmental treaty adopted on May 9, 1992, and opened for signature at the Earth Summit in Rio de Janeiro from June 3 to 14, 1992. It then entered into force on March 21, 1994, after a sufficient number of countries had ratified it.

(2) COVID-19

An *epidemic* is an outbreak of a disease that occurs over a wide geographic area and affects an exceptionally high proportion of the population. A *pandemic* is an epidemic of disease that has spread across a large region, for instance multiple continents, or worldwide. The 1918 influenza pandemic was fairly recent history. It was caused by an H1N1 virus with genes of avian origin. Although there is not universal consensus regarding where the virus originated, it spread worldwide during 1918–1919. On March 11, 2020, the World Health Organization officially declared the COVID-19 outbreak a *pandemic* due to the global spread and severity of the disease.

From Epidemic to Pandemic

COVID-19 illustrates how an epidemic becomes a pandemic. A mild to severe respiratory illness that is caused by coronavirus (of the genus *Betacoronavirus*) is transmitted chiefly by contact with infectious material (such as respiratory droplets), and is characterized especially by fever, cough, and shortness of breath and may progress to pneumonia and respiratory failure. The best way to prevent and slow down transmission is to protect yourself and others from infection by wearing a mask, keeping social distance, and practicing upscaled hygiene, including frequent hand-washing and not touching your face. Some people infected with the COVID-19 virus will experience mild to moderate respiratory illness and recover without requiring special treatment. Others develop long-term damage to major organs. Older people, and those with underlying medical problems like cardiovascular disease, diabetes, chronic respiratory disease, and cancer, are more likely to develop serious, and sometimes fatal, illness. The US death rate is, in fact, staggering, due, in large part, to the skyrocketing costs of healthcare and the impoverished conditions in which the American poor live. The national death rate of those who are infected and show serious symptoms is 1.3 percent, compared with 0.1 percent who do not survive the flu. Since it is a novel virus, there is still much to learn about the short- and long-term effects on the body caused by infection as well as the level of protection and the duration of the protection provided by the antibodies the body develops in response to the virus. On March 11, 2020, the World Health Organization characterized the new coronavirus (COVID-19) as a pandemic.

Sources: BBC: https://www.bbc.com/news/science-environment-11833685; https://www.cdc.gov/; U.S. Global Change Research Program: https://www.globalchange.gov/climate-change/glossary.

INDEX

CPSIA information can be obtained
at www.ICGtesting.com
Printed in the USA
LVHW091122120222
710692LV00028B/114